Hydroponics

How to build a hydroponic system at home, grow vegetables, fruits and herbs

I0390106

Table of Content

Introduction

Hydroponics are necessarily soilless plants that grow. It is a more effective way to give your plants food and water. Plants don't need fertilizer-they're using the food and water in the fertilizer. Soil's function is to supply plants nutrients and to anchor the plants' roots. You provide your plants in a hydroponic garden with a complete nutrient formula and an inert growing medium to anchor the roots of your plants so they can have easier access to the food and water.

It goes directly to the roots because the food is dissolved in water. Plants grow faster and are quicker to harvest. With a soil garden, you can grow more plants in the same area as you can, and because there is no soil, there is no concern about soil-borne diseases or pests-and no weeding.

Hydroponics is a method of growing soilless crops. Like in a typical greenhouse, plants are grown in rows or trellises, but they have their roots in water rather than soil. Many of us associate the nutrients with soil. In reality, the soil is providing support for plant roots, not the actual food itself.

The food comes from other soil-mixed products, including compost, broken-down plant waste, or fertilizers. Plants grown hydroponically will potentially grow faster and healthier than plants in the soil because they don't have to combat soil-borne diseases; besides, all the food and water they need is provided to their roots right around the clock.

Hydroponically, growing plants needn't be achieved on a wide scale, so it's simpler than you would imagine. Kits do-it-yourself systems and even fully automated growing tables are now available, all built for home gardeners. Hydroponics is really simple – it's easier in several respects than growing plants in soil. Plants need nourishment, water, and air. When you break it down to those three things, giving plants just what they need becomes easy. Hydroponics is the soilless method of developing plants.

The plants survive independently on the nutrient solution; the medium serves merely as food for the plants and their root systems. Hydroponics is basically soilless plants that grow. It is a more efficient way to give your plants food and water.

Plants don't need fertilizer-they're using the food and water in the fertilizer. The role of soil is to supply nutrients for plants and to anchor the roots of the plants. You supply your plants in a hydroponic garden with a full nutrient solution and an inert growing medium to anchor the roots of your plants so they can have better access to the food and water.

It goes straight to the roots since the food is dissolved in water. Plants grow faster and are quicker to harvest. With a soil garden, you can grow more plants in the same room as you can, and as there is no soil, there is no concern about soil-borne diseases or pests-and no weeding.

"I think it could allow us to travel farther and be more comfortable, whether that's underwater or above atmosphere. " Erik Bisa

Chapter 1: Introduction to Hydroponic Farming

Hydroponic agriculture is a method of growing plants using solutions of mineral nutrients, in water, without soil. The hydroponic gardener controls nutrient composition in the liquid solution that is used to water the plants. He/she also controls the amount the plants are supplied with nutrients. The hydroponic gardener manages the plants growing environment. Naturally, the device is highly automated but still needs to be handled well. As previously mentioned, the process is regulated and not operated merely. This is also water-efficient and nutrient-efficient, all of which are supplied straight to the root structure of the plant. Since the water and nutrient levels are tracked, these elements are provided at the appropriate levels as and when needed. Water and nutrients, together, contribute to success and growth rate.

The luminance element in crop production is also important. It is accomplished by planting out into vertical structures where illumination is maximized while reducing plant height, crowding, and shading.

Hydroponic farming of the present-day follows the 3-D method and is grown vertically in multilevel growing beds.

The potato was said to be domesticated between 8000 BC and 5000 BC in the South American Andes, along with rice, coca, llamas, alpacas, and guinea pigs. At the same time in Papua New Guinea, bananas were cultivated and hybridized. In Mesoamerica, 4000 BC domesticated wild teosinte to maize. Cotton was domesticated by 3600 BC in Peru. Camels were late domesticated, maybe some 3000 BC.

The Bronze Age, in c. 3300 BC, experienced the intensification of agriculture in cultures such as Mesopotamian Sumer, ancient Egypt, Indian subcontinent Indus Valley Civilization, ancient China and ancient Greece. During the Iron Age and Classical Antiquity period, the expansion of ancient Rome, first the Republic and then the Empire, in the ancient Mediterranean and Western Europe built on established agricultural systems while also creating the manorial structure that became the foundation stone of medieval farming.

Agriculture was transformed in the Middle Ages, both in the Islamic world and in Europe, with advanced methods and the spread of crop plants, including the introduction of sugar, rice, cotton, and fruit trees such as orange trees from Al-Andalus into Europe.

Following Christopher Columbus 'journeys in 1492, the Columbian trade brought Europe New World crops such as maize, onions, sweet potatoes, and cassava, and Old World crops such as wheat, barley, rice, and turnips, and livestock including horses, cattle, pigs, and goats.

Irrigation, crop rotation, and fertilizers were introduced shortly after the Neolithic Revolution, and grew over the past 200 years, beginning with the British Agricultural Revolution. After 1900, in the developed countries, and to a lesser degree in the developing world, agriculture has seen significant productivity increases as human labor has been replaced by mechanization and aided by synthetic fertilizers, pesticides, and selective breeding. The Haber-Bosch process permitted the industrial-scale synthesis of ammonium nitrate fertilizer, which significantly increased crop yields. Modern farming has raised financial, political, and environmental issues, including water contamination, biofuels, genetically modified organisms, tariffs, and farm subsidies. To respond to this, organic farming evolved as an alternative to the use of synthetic pesticides in the 20th century.

Original farmers have produced crops and animals which have flourished and thrived in various environments. They developed adaptations in the process to conserve soils, ward off frost and freeze cycles and protect their crops from animals.

Chinampa Wetland Farming

The Chinampa Field System is a form of elevated field farming ideally suited to wetlands and lake margins. Chinampas are built using a network of channels and narrow fields, built up and refreshed from the muck of the organic-rich river. Chinampas were used as long as 1000 BCE in the Lake Titicaca region of Bolivia and Peru, a method that supported the great civilization of Tiwanaku. The chinampas had fallen out of use about the time of the Spanish invasion of the 16th century. In this interview, Clark Erickson explains his experimental archaeology project, in which he and his colleagues engaged the Titicaca region's local communities to recreate elevated fields.

Mixed cropping

Though monocultural fields like this wheat field in the state of Washington are lovely and simple to manage, they are prone to crop diseases, infestations, and droughts without the use of applied chemicals.

Mixed cropping, also known as intercropping or co-cultivation, is a form of agriculture involving the simultaneous planting of two or more plants in the same field. Unlike today's monocultural systems (illustrated in the picture), intercropping provides a variety of advantages, including natural resistance to crop diseases, infestations, and droughts.

The Three Sisters

The Three Sisters, is a form of mixed crop system where in the same garden maize, beans and squash were grown together. The three seeds planted together, with the maize serving as support for the beans, both are acting as the squash's shade and humidity control, and the squash is acting as a weed suppressant. Recent scientific work, however, has shown that the Three Sisters were useful beyond that in quite a few ways.

Ancient farming technique: Slash and Burn Agriculture

Slash and burn agriculture – also known as swidden or shifting agriculture – is a traditional method of tendering domesticated crops that involves the rotation of many parcels of land in a planting period. Swidden has its critics, but when used with proper timing, it can be a reliable system for allowing the soil to be regenerated during the fallow periods.

Viking Era Landnám

We can learn a great deal from past mistakes too. When the Vikings founded farms in Iceland and Greenland in the 9th and 10th centuries, they used the same practices in Scandinavia they had used at home. Direct transplantation of ineffective farming practices is commonly seen as responsible for Iceland's environmental deterioration and, to a lesser extent, Greenland.

Significant numbers of grazing livestock, cattle, sheep, goats, pigs, and horses were brought by Norse farmers practicing landnám (an Old Norse term loosely translated as "land take." The Norse transferred their livestock to summer pastures from May to September, as they had done in Scandinavia, and to individual farms during the winters. To establish pastures, they cleared stands of trees, and cut peat and drained bogs to irrigate their fields.

Unfortunately, unlike the Norway and Sweden soils, the Iceland and Greenland soils are produced from volcanic eruptions. They are silt-sized and relatively low in clay, with high organic content and much more prone to erosion. In destroying peat bogs, the Norse decreased the number of native plant species adapted to native soils, and often competed with other plants and squeezed them out.

Extensive manuring in the first few years after settlement helped to improve the poor soils, but after that, and while the number and variety of livestock decreased over the decades, the deterioration of the climate went worse.

The problem was compounded by the beginning of the Medieval Little Ice Age from around 1100–1300 CE, when temperatures fell dramatically, impacted land, livestock, and people's ability to live. Finally, the colonies collapsed on Greenland.

Measured Damage Recent environmental damage estimates in Iceland suggest that since the 9th century, at least 40 percent of the topsoil has been destroyed. Soil erosion has affected a staggering 73 percent of Iceland, and 16.2 percent of that is categorized as bad or very extreme. 90 of the 400 known species of plants in the Faroe Islands are Viking-era imports.

Core Concept: Horticulture

Horticulture is the formal name of the ancient discipline of cultivating a garden crop. The gardener prepares the soil plot to plant seeds, tubers, or cuttings, helps to control the weeds and defends it against pests from animals and humans.

Garden crops are harvested, processed, and typically stored in containers or structures of specialized use. Some produce, sometimes, a large portion may be consumed during the growing season, but the ability of food storage for future consumption, trade, or ceremonies is an essential element in horticulture.

Maintaining a garden, which is a more or less permanent site, requires the gardener to remain nearby. Garden production has value, and a community of people will collaborate to the degree that they can defend themselves and their goods from those who steal it. Many of the earliest wildlife farmers even worked in fortified villages.

Archeological evidence for horticultural practices includes storage pits, tools such as hoes and sickles, residues of plants on specific devices, and improvements in plant biology that contribute to domestication.

Key Concept: Pastoralism

Images Pastoralism is what we term animal herding – whether it's goats, sheep, horses, camels or llamas. In Near East or Southern Anatolia, pastoralism was developed alongside agriculture.

Core Definition: Seasonality

Seasonality is a term used by archeologists to explain what time of year a particular site was occupied or other behavior. It is part of ancient farming, as people in the past organized their actions around the seasons of the year just as they do today.

Core Concept: Sedentism

Sedentary is the settling process. One of the consequences of relying on plants and animals is that social care is needed for those plants and animals. One of the reasons historians often claim that humans were domesticated at the same time as animals and plants are the behavioral shifts in which humans build houses and live in the same places to tend crops or take care of livestock.

Core Concept: Subsistence

Subsistence refers to the collection of new activities that humans use to get their food, such as hunting animals or birds, fishing, collecting or tending plants, and full-fledged agriculture.

The milestones of human survival evolution include sometimes fire control in the Lower to Middle Paleolithic (100,000-200,000 years ago), game hunting with stone projectiles in the Middle Paleolithic (about 150,000-40,000 years ago), and food storage and an increasing diet by the Upper Paleolithic (about 40,000-10,000 years ago).

Agriculture was discovered about 10,000-5,000 years ago in various locations in our country, at different times. Scientists study historical and prehistoric agriculture and diet using a wide variety of objects and measurements, including Forms of stone tools used to process food, such as grinding stones and scrapers Remain of storage or cache pits that include small pieces of bone or vegetable matter Middens. These garbage waste deposits include bones or plant matter.

Microscopic plant residues sticking to the edges or faces of stone tools such as pollen, phytoliths and starches Stable animal and human bone isotope analysis Dairy Farming is the next step after animal domestication: people keep cattle, goats, pigs, horses and camels for the milk and milk items they can supply. Archeologists, once regarded as part of the Secondary Products Revolution, are coming to agree that dairy farming was a very early phase of agricultural innovation.

A midden is simply a garbage dump

Archeologists love middens because they also have knowledge about diets and the plants and animals that fed the people who used them that is not otherwise available.

Eastern Agricultural Complex, The Eastern Agricultural Complex, refers to the variety of plants selectively tended by Native Americans in eastern North America and the American midwest, such as sump weed (Iva annua), goosefoot (Chenopodium berlandieri), sunflower (Helianthus annuus), small barley (Hordeum pusillum), erect knotweed (Polygonum erectum).

1.2 Role of hydroponics in farming

Hydroponics

It is a subset of hydroculture and is a method of growing soilless plants by using mineral nutrient solutions in a water solvent. Terrestrial plants can be built with only their roots exposed to the nutritious liquid, or an inert surface such as perlite, gravel, or other substrates can physically support the sources. Given inert media, sources can induce changes in the pH of the rhizosphere and root exudates can influence the biology of the rhizosphere.

These in between nutrients that are used in hydroponic systems can come from a number of sources like, but that are not limited to, fish excrement, duck manure, purchased chemical fertilizers, or artificial nutrient solutions.

Tomatoes, peppers, cucumbers, lettuces, marijuana, and model plants such as Arabidopsis thaliana are commonly grown hydroponically on the inert paper.

Hydroponics provides many advantages, one of which is a decrease in water use for agriculture. Growing 1 kilogram of tomatoes in intensive farming requires 400 liters of water, 70 liters of water in hydroponics, and just 20 liters of water for aeroponics. Due to the scarcity of water to produce, it will be feasible in the future to grow their food for harsh conditions that do not have much available water.

Here is a list of things showing the value of hydroponics:

- Crops Rising In Any Environment
- Lower Pesticide Consumption
- Higher Food Production
- Water Quality
- Vertical Farming
- Addressing the World Hunger

No doubt, hydroponics will have a positive effect on the future. Citizens have long been neglecting hydroponics. Do we doubt if hydroponics is worth it?

With the increasing rise in global hunger and climate change, the food supplies of the earth will dwindle gradually. I think hydroponics will play an essential role in overcoming the next challenges in the future.

Growing crops

Hydroponic crops grow in a controlled environment at any temperature. Factors such as temperature, water supply, and sunlight are important to plant survival. Due to the season, these variables are continuously changing during the year. When using hydroponics for planting crops, the weather variations do not impact the plants. Farmers can thus grow vegetables throughout the year and increase their yield by up to three times as much.

Throughout winter, crops such as corn and tomatoes are hard to grow before using hydroponics. Yet thanks to this new farming approach, high yield will be available throughout the year.

As a result, because of the constant supply current, there would be no demand fluctuations. There are also areas where other crops grow by cant.

There's no question that the country's geographical position will influence its environment. Some states have temperatures either too high or too low that they can't cultivate those crops.

States such as Alaska have temperatures that can often fall to -15 ° C. Temperatures below freezing, right? Can't grow rice, wheat, or even tomatoes in areas like this. Farmers will grow carrots, broccoli, and cabbage there. But there grow strategic crops like wheat and rice cant without the use of hydroponics and a greenhouse. We don't want you to believe that temperature is the only influencing factor in farming, after considering all of the previous examples regarding climate. There are places where ideal temperatures and a sunny daytime are available, but there is another crucial element missing, that is soil quality.

Poor soil quality can cause significant damage to any crop to be cultivated. Many factors decide the quality of the soil, but the most important are the nutrients present and PH. Having low PH soil can significantly affect the ability of your plant to take up nutrients. Additionally, early nutrient deficiency syndromes can occur on your plants when the land you are planting in has low nutrient levels. These syndromes can cause stunted growth and low final production.

Lower use of pesticide

Plants is grown in the open air in soil farming, which makes them more vulnerable to different pests. Such pests may come through several methods to your crops. Many pests spread from neighboring farms, and other pests from far-away farms get carried by the wind. Pests the hydroponics, on the other hand, have some closed system. Most hydroponics are grown in greenhouses. Those greenhouses serve as a buffer against infection with pests.

Infections with pests can, of course, occur in hydroponics farming. These events, however, are rare to occur. Pests and insects can pass through wall holes or ventilation systems, and other pests get into the greenhouse by simply sticking to the farmer's clothes before they enter. Almost all hydroponic farms have standard protection measures to avoid any contamination with the disease. Another concern soil brings with it is marijuana.

Weed is any unwanted plant that grows alongside your crops. You may think that having unwanted plants next to your crop is not a big deal, but be vigilant about its vast consequences. Weed and your plant compete on different growth factors.

You would be shocked by the stunted growth of your plant due to the grass growing alongside it. Weed uses the same supply of water as your crops, so it decreases the number of nutrients in your soil. Nutrient deficiency syndromes should begin setting in by the end of the rising process.

Weed compete on diverse growth with your plant Hydroponics solves the question of weed. With Rockwool, your plants should grow up in net pots. Weed can't grow and compete with your plants because it can't grow. Also, because of the greenhouse, weed seeds that get transported by wind would not be able to touch your crops.

All these properties protect the plants in hydroponic conditions. Pesticides or weed control products are not required. Such goods are harmful and have numerous adverse impacts on the human body. The creation of a pesticide-free crop is one of the most important reasons why hydroponics should be our primary method of cultivation.

Higher Food Production,

Almost all scientific work on hydroponics, has found one significant fact; plants with hydroponics tend to have a higher growth rate and final yield than conventional plants with soils.

Plants expend a large amount of energy on root growth in the soil; this is a natural strategy of survival that plants do to access more water and nutrient supplies.

Hydroponic plants, on the other hand, are not spending energy on root growth. Plant roots have easy access to the nutrient solution in hydroponics. Therefore the plant doesn't need to waste its resources on root expansion. Instead, this energy saved is used to improve the growth rate. Consequently, hydroponic plants yield more food than typical soil plants. Soil plants continue to be influenced by normal conditions at the site. Factors such as sunlight and temperature change slowly during the day. Yet, in a hydroponic greenhouse, these factors are highly regulated. The temperature and sunlight are monitored regularly to ensure optimal plant conditions. Additionally, concentrations of carbon dioxide are managed by specific equipment in advanced facilities to allow the photosynthesis process to run smoothly.

A research was performed to compare the lettuce in soil and hydroponics. To complete the growing process, the plants are kept for 40 days. The findings were in favor of hydroponics by keeping all the variables like sunlight and concentration of nutrients stable. Hydroponic lettuce has grown to 160 g fresh mass relative to soil lettuce, which only grew to 120 g.

Water Quality

Most people believe that the roots get most of the water when the water from their soil plants, which is untrue. Soil plants get a tiny portion of the water applied to the soil. And where is the leftover water going? Much of the water that is applied to the soil get lost by various methods. A substantial portion of water gets lost from the soil surface through evaporation, and another portion gets lost to the ground by drainage. If weed is next to the plants, there is a strong probability that it will absorb a good portion of water, too.

The soil has reduced water capacity, but hydroponics demonstrates promising efficient use of water. In hydroponics, the water that evaporates is almost 0 percent. Evaporation does not occur because the surface of the water does not receive so much heat as soil. The only route of dehydration could occur only through the leaves of the plant.

There's still no water that's lost to the table. Almost everything of the water is being reused. If nutrient levels go down, nutrient solutions may be applied back to the water without the need for a substitute.

For this reduction of water usage, there would be a significant likelihood that hydroponics will help save the world.

We assume that hydroponics can play a crucial role in supplying food for the world's population. It not only uses 95 percent less water than soil but also provides a higher yield than soil.

Vertical farming

The combination of vertical farming and hydroponics would increase the final yield considerably. Vertical hydroponics has provided more return per m2 compared to soil.

If you grow on the field, then you will be able to use the growing area for once. Imagine how much yield you'll achieve when you've used several times the same growing field. Vertical hydroponics allows increasing the yield by increasing different hydroponic rates above each other. But instead of only using the growing area for once, you can use it more than ten times, depending on how much rates have increased. With the expansion of vertical hydroponics, more vertical farms present amid crowded cities would be fair to see. Farms can no longer be contained in rural areas.

Farmers continue to migrate to rural areas because of the low cost of the land they have; this has resulted in transport costs being added to the price of the yield.

The result is that the customer must also pay for delivery. There will be no transportation added by having vertical farms in the center of crowded towns, as the farms are nearly beside most customers.

Food security and global hunger

Nearly 736 million people live in extreme poverty for many reasons, according to the World Bank Organization. The key explanation for this is continuously rising food prices.

By combining conventional soil farming with hydroponics, there is a strong possibility of addressing the problem of world hunger. Hydroponics not only yields more than soil, but it can also be a cheaper process.

The incorporation of solar systems and automated hydroponic devices would dramatically reduce their costs. As a result, food prices will plummet, and more people will be raised above the poverty line.

1.3 Types of hydroponics farming

Six common forms of hydroponic systems exist:

- Wick
- Water Culture
- Ebb and Flow (Flood & Drain)
- Drip (recovery or non-recovery)

- N.F.T. (Nutrient Film Technique)
- Aeroponic

Such basic types of systems are subject to hundreds of variations, but all hydroponic methods are a variant (or combination) of these six.

WICK

The Wick method is by far the simplest type of hydroponic system. This is a passive device, meaning that there are no moveable pieces. The nutrient solution is drawn in from the reservoir with a wick into the growing medium. There are free plans for a simple wick program (please click here for plans). This program can employ a variety of can media. Among the most common are Perlite, Vermiculite, Pro-Mix, and Coconut Fibre. The system's greatest downside is that plants that are massive or require large quantities of water can consume the nutrient solution quicker than the wick(s) can supply it.

WATER CULTURE

For all active hydroponic systems, the water culture method is the simplest. Typically the frame holding the plants is made of Styrofoam and floats directly on the nutrient solution. An air pump provides air to the air stone, which bubbles the nutrient solution and supplies the plant roots with oxygen.

Water culture is the preferred method for growing leaf lettuce, which is rapidly increasing water-loving plants, making them an excellent choice for this very type of hydroponic system. In this type of policy, very few plants other than lettuce should do well.

For the classroom, this form of hydroponic system is excellent and popular with teachers. It is possible to make a very inexpensive device out of an old aquarium or another watertight container. We have free plans and guidance for a basic water culture program. The only downside of this method is that it doesn't work well with huge plants or long-term plants.

EBB & FLOW - (FLOOD AND DRAIN)

The Ebb and Flow mechanism work by flooding the rising tray temporarily with a nutrient solution, and then it works by draining the solution back into the reservoir. Typically this operation is performed with a submerged pump that is attached to a timer.

When the timer clicks, the pump is pumped into the rising tray on a nutrient solution. When the speed timer turns the pump off, the nutrient solution flows back into the tank. Depending on the size as well as a variety of plants, temperature and humidity, and the growing medium used, the timer is set to turn on multiple times a day.

The Ebb & Flow method is flexible and can be used with a variety of that media. Grow Rocks, gravel, or granular Rockwool will fill the entire grow tray. Many people like to use individual pots generally filled with growing medium, which makes it easier to move or even move plants inside or out of the network. The critical drawback of this kind of device is that there is a susceptibility to power outages as also pump and timer failures with certain forms of growing medium (Gravel, Growrocks, Perlite). The roots will dry out quickly on the interruption of the watering cycles. Through using that media that retains more water

DRIP SYSTEMS RECOVERY / NON-RECOVERY

Drip systems are possibly the most commonly used form of hydroponic system in the world; this problem can be somewhat alleviated. Operation is simple; a timer regulates a submersed pump. The timer turns the pump on, and a short drip line drips nutrient solution onto each plant's foundation. The excess nutrient solution which runs off is collected back in the reservoir for reuse in a Recovery Drip System. The Non-Recovery Program receives no runoff.

A recovery system uses a little more effectively the nutrient solution because the excess solvent is reused, this often enables the use of a cheaper timer as a recovery system does not require precise monitoring of the watering cycles. The non-recovery system requires a more reliable timer so that irrigation cycles can be modified to ensure that plants obtain ample nutrient solution and that the runoff is kept to a minimum.

The non-recovery method needs less maintenance because the surplus nutrient solution isn't pumped back into the reservoir, so the reservoir's nutrient intensity and pH won't differ. This means you can fill the tank with pH-adjusted nutrient solution and then forget it until more mixing is needed. A recovery system can have significant changes in the levels of pH and nutrient intensity that involve periodic checks and adjustments.

N.F.T. (Nutrient Film Technique)

What most people think of as a hydroponic device when they think about hydroponics? N.F.T. systems provide a continuous flow of nutrient solution, so no timer for the submersible pump is needed. The nutrient solution is gradually pumped into the growing tray (usually a tube) and flows over the plant's roots, then drains back into the tank.

Usually, there is no growing medium used other than air, which saves the cost of replacing the growing medium after each crop. The plant is generally protected in a small plastic basket where the roots hang in the nutrient solution. N.F.T. systems are particularly vulnerable to power outages and pump failures. When the flow of nutrient solution is disturbed, the roots dry out very quickly.

AEROPONIC

Potentially the most high-tech form of hydroponic gardening, aeroponic device. As above through a medium, the N.F.T. device is mainly air. The roots remain in the air and are misted with a solution of nutrients. In general, the misting's are done every few minutes since the roots are exposed to the air like the N.F.T. method. If the misting cycles are disrupted, the roots can dry out quickly.

A timer controls the nutrient pump in much the same way as other forms of hydroponic systems, but the aeroponic system needs a short cycle timer running the shoe every few minutes for a few seconds.

1.4 Is traditional farming same as hydroponics farming?

There is nothing more like a bit of competition, and the grocery store is ground zero for food growers. Hydroponics hit the ground running since entering popular culture just a couple of decades ago, proving to be a formidable competitor to conventional farming behemoths.

Is this latest comer zippy, sterile, and high-tech any match for a century-old way of living? When you look at what's become of modern agriculture, the reaction is a resounding yes.

Contemporary agriculture is no longer a family-owned home business kept together by the rural family. Big Ag and mass plantings are the manners in which the agriculture industry earns much of its income. The durability of those activities is uncertain. But hydroponics what? RosebudMag.com weighs in on seeing who's taking the cake.

1. WIN-Hydroponics: Lowest use of pesticides.

Hydroponic produce is grown under sterile indoor conditions that prevent soil-borne pests and disease from being introduced.

It has its ups and downs because if one should arise, growing several plants in an exceptionally small area can lead to a widespread and difficult outbreak to manage. But hydroponic pest control is mainly focused on preventive measures, and any successful grower should keep their crop sterile by ensuring proper cleaning of all that goes into the growing room.

If mischievous creatures find their way into a closed hydroponic system, most farmers prefer organic pest control methods such as ladybugs, copper wire, neem oil, ammonia, pyrethrin, insecticidal soap, and other natural substances. If (and this is if) synthetic pesticides were used on a hydroponic crop, their effect would still be hypothetically lower than that of a conventional plant, as the chemicals would be trapped and unable to leach directly into the soil or ultimately find their way into water tables.

Although soil-based agriculture is capable of earth-friendly management of pests like the kind used by the organic industry, most farmers tend to douse their crops with any amount of chemicals to ensure that anything within a mile radius drops dead.

2. Impact of WIN-Hydroponics use of fertilizers.

Since hydroponics is a closed system, fertilizers used to feed crops are cycled repeatedly when irrigating the plants. When a certain nutrient balance is achieved in the reservoir, this water can be stored anywhere from 10 to 14 days before its nutritional content is exhausted. Still, the solution is available after that. This can be uniformly sprayed over rising plants in the field to give them a little lift, although diluted.

Soil-based agriculture is a fertilizer glutton, at least to what it has become today. Farmers flood their fields with both inorganic and organic fertilizers at close intervals to ensure crops have much more than they need to grow their absolute best.

This collective overreaction has catastrophic consequences for water supplies and the ocean. Since the 1980s, unsustainable quantities of manure from Missouri, Arkansas, Mississippi, Tennessee, and Ohio have poured into the Gulf of Mexico, leaving the southern US coastline a virtual dead zone.

Because of this year, because of the algae blooming effects, a region the size of New Jersey is devoid of life.

When the fertilizer penetrates the sea, colonies of algae erupt and die rapidly. A decomposition chokes the oxygen water and marine life, is no longer able to breathe, either dies with it, or is driven out of the area.

3. DISPUTED (but most likely conventional agriculture wins) Lowest energy consumption.

It is a hot subject that places hydroponics on the powder keg: Does its energy consumption outweigh the benefits?

Hydroponics could be a complete energy sucker if growers don't use the most up-to-date lighting technologies available. Yet there's still space to grow in a greenhouse or even outdoors. If the power of the sun is harnessed by a hydroponic farmer and/or augmented by high-efficiency growing lights, their operation is sound.

But what if the plug gets pulled or people have to turn back to the land? If times were very hard, a hydroponic set-up could be managed with a lot of hand watering and continuous monitoring of the power.

That is where farming can take pride in itself. Outdoor soil-based crops harvest the sun's free energy, one of the few aspects that they have still not been able to kill.

That's the reason that human beings have made that where they are now, from the cradle of civilization to mass food production - seeds sown in the earth, fed, and left under the sun have forever changed the path of humanity.

4. WIN-Hydroponics: Lowest water use.

Hydroponics is known for using up to 90 percent less water than conventional farming because of its closed system design. Water is cycled over and then over again to irrigate fields, making it suitable for drought situations or places with limited access to freshwater, such as deserts (i.e., Nevada) or radiation zones (i.e., Fukushima), for a highly efficient method.

Agriculture relies on irrigation, which is fine, but when there is a drought, this can also lead to poor yields and puts farmers at the mercy of the environment. This way too, arable land is limited, since, in some regions, it is impossible to locate a water source that is reliable enough to irrigate a field.

5. Most lucrative WIN - Hydroponics for a home on the coast.

Eventually, hydroponics may be credited with reviving the family farm.

If a landowner has a small or unsuitable piece of land, rows, and rows of plants may still be used to grow on top of each other in a greenhouse.

Not only does this optimize rising capacity, but it also offers small producers a chance to break into a market with a hydroponic niche that would otherwise dominate.

And because many leafy crops such as lettuce and herbs have a very short processing period from seed to harvest, the yields themselves can be higher than if they were planted in soil, which is all equal to a tidy business for the new farmers.

For the most part, agricultural companies dominate agriculture, but small-scale family-run farmers and organic farmers (bless their souls) are alive and thriving too, albeit certainly a minority.

What wins, then? When you look at these five things, it's as if hydroponics is taking the throne. But the reality is both hydroponics and conventional farming are required to ensure global food security. Agriculture has a lot of clean-up to do, and there will be hydroponics to fill up the gaps ... and maybe even take over one day.

1.5 Benefits of Hydroponic farming

1. No soils required

In a sense, in areas where the soil is poor, does not exist, or is heavily polluted, you can grow cultivated. Hydroponics was widely used in the 1940s to supply fresh vegetables for troops in Wake Island, a Pan American airline refueling stop. It is a small region of Pacific Ocean arable. Hydroponics has also been regarded by NASA as the potential farming to grow food for astronauts in space (where there is no soil).

2. Make better use of space and location

As all that plants need is given and maintained in a system, you can develop in your small apartment, or the spare bedrooms as long as you have some space.

The roots of plants typically grow and spread in search of food, and in soil, oxygen. For Hydroponics, where the roots are submerged in a tank full of oxygenated nutrient solution and close contact with essential minerals, this is not so. This means you can grow far closer to your plants, and ultimately save huge energy.

3. Climate control

As in the greenhouses, hydroponic growers may have complete climate control-temperature, humidity, light intensification, air composition.

In that sense, no matter the season, you can grow food all year round. Farmers should grow food at the right time to maximize income for their businesses.

4. Hydroponics is water-saving

Plants that are grown hydroponically will only use 10 percent of water compared to those grown on the ground. Water is recirculated using this form. Plants will take the water they need, while the run-off ones will be caught and returned to the network. Water loss occurs only in two ways-evaporation and device leaks (but a successful hydroponic design will reduce or have no leaks).

Agriculture is expected to use up to 80 percent of ground and surface water in the United States.

Although water will become a critical problem in the future as the FAQ predicts that food production will increase by 70 percent, Hydroponics is considered a feasible option for large-scale food production.

5. Efficient nutrient usage

In Hydroponics, you have 100 percent control over the nutrients (foods) plants need. Before planting, growers should test what plants need and how much nutrients they need at different levels, and combine them with water accordingly.

Nutrients are safely conserved in the tank, and there are no nutrient losses or shifts as they are in the soil.

6. Solution regulation pH

All minerals are found in the solution. This ensures this compared to soils; you can calculate and change the pH levels of your water mixture much better. That ensures the maximum absorption of plant nutrients.

7. Better growth rate

Are plants grow faster in hydroponic form than in soil? Indeed, it is. You are your own boss, who controls the entire ecosystem for the growth of your plants-temperature, light, moisture, and nutrients in particular. Plants are put in ideal conditions, while nutrients are supplied in adequate quantities, and the root systems come into direct contact. Thus, plants no longer spend precious energy checking the soil for diluted nutrients. Instead, they move their entire emphasis to grow and producing fruit.

8. No weeds.

You'll understand how annoying weeds affect your garden when you've worked in the soil. For gardeners, it is one of the most time-consuming activities-till, plow, hoe, etc. Weeds are mainly soil related. So clear the nutrients, and all weed bodies are gone.

9. Fewer pests & diseases

Just like weeds, getting rids of soil helps make the plants less susceptible to soil-borne pests like birds, gophers, groundhogs, only diseases like the species Fusarium, Pythium, and Rhizoctonia. Even when growing indoors in a closed system, gardeners can easily take care of most environmental variables.

10. Less use of insecticide and herbicides

When you don't use soils, so although weeds, pests, so plant diseases are significantly reduced, fewer pesticides are used. It will help you eat safer and cleaner foods. Insecticide and herbicide cutting is a strong point of Hydroponics as the standards for real life, and food protection is gradually being put on top of it.

11. Labour and time savers

In addition to spending fewer hours on tilling, watering, planting, and fumigating weeds and pests, you enjoy saving a lot of time as in Hydroponics the growth of plants is known to be higher. Hydroponics has a space in it, as agriculture is expected to be more technology-based.

12. Hydroponics is a hobby that relieves stress.

This is a great interest that will bring you back in touch with nature. Tired after a hard day of work and traveling, you're back to your tiny corner of the apartment; it's time to lay it all back and play in your hydroponic garden. Reasons such as lack of spaces no longer make sense. You can fill your little closets with fresh, delicious vegetables, or essential herbs, and enjoy the relaxing time with your little green spaces.

Hydroponics sounds like there are tons of benefits, and the picture below appears to seek to convince you into rising Hydroponics. Yet start reading to hear about its downsides.

1.6 Applications of hydroponic farming

After all, people used good old-fashioned soil to cultivate food just fine for thousands of years, if not millions of years. Hydroponics provides some significant advantages over conventional farming, and as word spreads about these advantages, more people turn to hydroponics for their agricultural needs.

Second, hydroponics provides people with the opportunity to grow food in areas where conventional farming literally cannot be achieved.

Since decades hydroponics has been in use in areas of arid climates, such as Arizona and Israel. This science helps those living in these areas to enjoy developed locally and to increase their food production. Hydroponics is equally effective in dense urban areas, where land is at a premium. In Tokyo, hydroponics is used in place of traditional plant growth based on soil. In remote locations, such as Bermuda, hydroponics is also useful. With so little space available for planting, Bermudians have turned to hydroponic systems that usually take up about 20 percent of the land needed for crop growth. It helps the island's people to enjoy local produce year-round without the cost and delay of imports. Finally, hydroponics can benefit areas that don't receive consistent sunlight or warm weather. Locations such as Alaska and Russia, where growing seasons are shorter, use hydroponic greenhouses to regulate light and temperature to achieve higher crop yields.

We also need to consider the considerable environmental benefits of using hydroponics. Hydroponic systems generally require only about 10 percent of the water required by soil-based farming.

This is due to the fact that hydroponic systems allow water and nutrient solutions to be recycled and reused, and that no water is wasted.

It can impact places where water is scarce, such as the Middle East and parts of Africa, quite a bit. Similarly, hydroponics needs little to no pesticides and just around 25 percent of the soil-based plants 'nutrients and fertilizers available. This always not only reflects cost savings but also protects the atmosphere, as no contaminants are released into the air. Finally, we have to consider the impacts of transport on the environment. Given that hydroponics allows production to be grown locally and needs fewer areas to generally import their crops, both price and greenhouse gas emissions are reduced due to reduced transport requirements

Hydroponics then gives us the benefit of shorter harvest time. Plants grown in this manner have easy access to water and nutrients and are thus not required to establish extensive root systems so that they can obtain the nutrients they need. As conventional agriculture, this saves time and yields better, lusher plants in around half of the time.

So why not take over the hydroponics? This is due to many distinct inconveniences associated with these systems.

The first is the heavy investment in resources compared with soil cultivation.

Although hydroponics is typically much cheaper over time, establishing any kind of bigger system requires a substantial upfront cost. Next, there is the danger of power failure, which can cause pumps to stop crops from functioning and destroy them. Lastly, many people are afraid that hydroponics takes extensive know-how and research, although it is quite close to conventional gardening indeed. After all, in order to grow, plants rely on certain nutrients, and those nutrients do not change, no matter which method you use.

A hydroponic greenhouse system uses soilless media to expand, providing a higher yield of fresh produce for farmers. Hydroponics offers a cheaper, safer, and more sustainable way for many farmers to grow incredible crops in small and non-traditional growing spaces.

Hydroponics key benefits include improved plant production, high yield per plant per square foot, and fresh produce. Today, many types of hydroponically grown plants represent several different consumer segments, including farm stands, grocery stores, restaurants, manufacturing plants, and institutions. Hydroponic operations range in scale from small operations (less than 1,000 square feet) to large operations where several acres are planted.

CANNABIS PRODUCTION & HEMP PROPAGATION

Exact regulation of the environment is important for the efficient development of cannabis. Making sure that you have the right amount of light and the right quality will significantly affect your harvest. Our greenhouse systems can assist you with heating, cooling, CO 2, air circulation, and lighting power. We also work with iGrow to ensure that all of these devices, including from your mobile phone, are easy to manage! Depending on the actual size of your operation, we sell freestanding or gutter link type greenhouses.

HIGH TUNNEL

High Tunnels are a growers 'increasingly common phenomenon and an established crop production technology. The word "high tunnel" is a broadly described phrase used in greenhouses for growing fruits and vegetables, though some high tunnels are used for cutting flower growth. The distinction between a High Tunnel and a standard free- greenhouse is that high tunnels use Eastpoint, Northpoint, and Nor Easter frames to make them "high" tunnels, with longer ground posts.

SHADE STRUCTURES

These multi-purpose structures are simple to install and offer a variety of color and density shade cloth choices. Shade structures can protect and relieve plant stress and reduce the need for watering. Everything you need, including the frame, hardware, and shade cloth, is factored into the price.

EASTPOINT

The Eastpoint Series is known to be the best of both worlds, mixing power and flexibility at an affordable price to make it a perfect greenhouse for starters. The Eastpoint is ideal for growers who want a larger freestanding house to last but are looking for something simple and easy to build.

NORTHPOINT

The Northpoint Series is the ultimate in freestanding greenhouses, built as one of the original Rimol Greenhouses, and planned as both a growing and retail space for multifunctional use. The Northpoint is extremely resilient and can withstand the harsh loads of snow and wind seen in all parts of the world.

NOR'EASTER

The Nor'Easter greenhouse collection is the fastest available freestanding greenhouse.

The Northeast United States is notorious for experiencing extreme and volatile weather, and, as the name implies, the Nor'Easter greenhouse series will protect your crops in some of the most severe weather conditions. The Nor'Easter can be relied on to be reliable when you most need it and is constructed with a special truss support system that doubles the strength of each bow. Rimol Greenhouses are designed to last, and this will be illustrated by the rock-solid design of the Nor'Easter sequence greenhouse.

ROLLING THUNDERTM

The Rolling ThunderTM, the mobile greenhouse solution from Rimol, enables you to set up your growth anywhere you need it. This mobile greenhouse device uses a heavy-duty wheel with bearings connected to a specially built ground post at each set of hoops. The combination of wheel/ground post is seated on a rail, which allows the greenhouse to travel along with the desired growth areas. Rolling Thunder's architecture allows only two individuals to comfortably transfer a bigger greenhouse with an ordinary tractor and a smaller greenhouse.

FREESTANDING GREENHOUSES

Education and rehabilitation classes, correctional institutions, and homeowners have all discovered the importance of greenhouses.

These are solid, long-lasting structures with a 20-year polycarbonate cover that is durable. Polycarbonate is resistant to fire, and the structure of galvanized steel is fireproof, making the building nearly indestructible. Our professional members will show you how others found those greenhouses.

MATTERHORN

The Matterhorn stands alone as our greenhouse with the most stunning, rugged, and typically styled gout. Inspired by that majestic peak in the Swiss Alps that attracts intrepid climbers to risk their lives every summer scaling its rugged terrain and ridges, the Matterhorn greenhouse is a high-quality, solid, and attractive structure that is ideally suited for garden centers, farmers and schools in any North American environment. The Matterhorn can itself withstand heavy snow loads and wind loads, so that ensures that only the worst winter storms, so hurricanes will give you a decent night's sleep.

Chapter 2: Building a Home-Based Hydroponics System

Commercial food production at residential and small scale can take many forms. The most popular may be traditional home gardens that use natural soil. Still, a general interest in the growing of vegetables is not limited to those with suitable outdoor and in-ground locations. For other instances, a gardener may not have access to a plot of land, or the soil may be of such low quality that it is not a choice to grow in the field. Soilless production and hydroponics are generally options for many and allow for the development of small-scale vegetables where conventional gardens would be then impossible.

The growing systems and all the techniques involved in soilless growing will help those who grow their food in urban areas with limited spaces, a sunny patio, or a variety of other locations and situations.

Growing plants without using soil have been achieved for many years, but these methods 'research and experience continues to evolve and improve opportunities for commercial growers and gardeners alike.

The educational publication was prepared in a joint effort between the University of Tennessee Extension and the University of Florida Joint

Extension systems to provide knowledge and introduction primarily for gardeners, teachers, youth, and other non-commercial growers to these hydroponic practices and techniques.

Decide the position Place the hydroponic device in an enclosed space, such as your house's greenhouse or basement, or an outdoor patio or deck; The floor should be level to ensure the plants in the system have equal coverage of water and nutrients. When the system is installed outside, shield the system from the weather such as having a wind barrier and most frequently test the water levels due to water loss from evaporation. Bring the hydroponic device indoors during cold temperatures. If you place the device in your house's interior space, add grow lights to support the plants with additional lighting.

2.1 Basics of Home-Based Hydroponic System

Hydroponic systems and how to construct your hydroponic systems. We are not offering any goods or details, and we can remain impartial. We're not just publishing lies for factories intended to market products like other websites. We strawberries are growing in a hydroponic home build system use only knowledge from reliable and creditworthy sources, as well as input from actual growers. Not only home growers but also what we are studying to shape commercial hydroponic farms.

Throughout the years, we've spent a lot of time looking for right creditworthy information to grow our plants, and Home Hydro Systems was developed to share what we've learned and figured out with everyone else as well. Simply because we want to help others successfully grow their plants, too. Around the same time, we will help to promote the benefits of using hydroponics to grow food. We generally have only one goal in mind, and that's to help show people that you can hydroponically improve your diet, and without it costing a lot.

Flowers on the strawberry plants in a hydroponic system The good growth of plants is not enough, it has to be affordable too, or what's the point. We're not interested in growing a $12 tomato; we're trying to show you that you can hydroponically grow better produce at home, and for less money than you can buy it at the supermarket, or even build it on the farm. After all, if it weren't economical to raise them hydroponically now, would it not be worth it?

Homemade hydroponic system growing lettuce you don't need a costly hydroponic system, expensive nutrients, or even expensive growing lights to cultivate your crops hydroponically.

You can create your own inexpensively very strong hydroponic systems. There's also plenty of really fine commercially generated cost-nutrients you can use. When you can make use of the abundant natural sunshine, you don't even need any grow lamps. Construction of your hydroponic systems is easy. Anything you want to develop, hydroponic systems only have and need a few simple parts to function well.

The hydroponic system needs only a few simple parts to build

Growing Chamber

The growing chamber is said to be the part of the hydroponic system where the plant roots grow. The rising room, to put it, is the root zone container. This region offers plant nutrition and is where the roots obtain the nutrient solution. It protects the roots against sun, heat, and pests, too. Keeping the root zone cool and light-proof, is essential. Prolonged light will affect the roots, and as a result of heat stress, high times in the root zone can cause heat stress to your plants, as well as cause fruit and flower drops. The temperature of the nutrient solution itself is a very critical part of keeping the roots, and the entire root zone is said to be comfortable for the plants.

The size and shape of all the growing chamber depend on your building's type of hydroponic system, as well as on the type of plants you would grow in it. Wider plants have wider root systems, so they require more room to live in. The designs are infinite here. You can use almost anything as the growing chamber, don't want to use something made of metal, or it may destroy or react with the nutrients. When you look around, you can get plenty of insights into what and how you can easily use several different things to construct your hydroponic system's growing chamber.

Reservoir

The reservoir is the component that contains the nutrient solution in the hydroponic system. The nutrient solution is composed of plant nutrients combined in water. Based on the form of hydroponic device, the nutrient solution that can be pumped from the saved reservoir up to the growing chamber (root zone) in cycles using a timer as well as continuously without a timer, or the roots can also hang down into the reservoir 24/7, making the reservoir the growing chamber.

With just about everything plastic that retains water, you can make a reservoir. This can be used as a tank as long as it does not leak, retains enough water and is washed out properly first. Read this book for more about how high your pool of nutrients will be. Likewise, a reservoir must be light proof. It's not light-proof, because you can keep it over your head and see the light flowing through it. But painting it, covering it, or wrapping something like bubble wrap insulation around it is easy to make any container light proof. With even low light levels, algae and microorganisms will start developing.

Submersible Pump

Most hydroponic systems pump water (nutrient solution) from the reservoir up to the growing chamber/root zone for the plants using a submersible pump. Submersible pumps can be found conveniently in the hydroponic supply shop, or most home improvement stores with garden supplies as pumps for fountains and ponds. They should come in a wide variety of sizes too. Read this page for determining what pump size you need for your hydroponic system?

The submersible pumps are nothing more than an impeller that spins it using an electromagnet. They can also be quickly put away to be thoroughly washed. If it doesn't come with a filter, cutting a piece of furnace filter screen or similar material to match will easily make one. To keep them clean, you should periodically clean both the pump and filter.

Delivery system

A water/nutrient solution delivery system for hydroponic systems is very easy and extremely customizable when constructing your hydro system. In addition to the pump, it's nothing more than just plumbing through the water/nutrient solution to get to the plant roots in the growing chamber and back to the reservoir. A mixture of standard PVC tubing and connectors, modern garden irrigation tubing, and connectors, as well as blue or black vinyl tubing, are usually the easiest and safest materials to use for the nutrient delivery system.

One wants to use drip emitters or sprayers as part of your nutrient solution distribution system, always depends on the type of hydroponic system you create. While useful, they can also clog. And if you do, make sure you've got extras that you can change out easily when cleaning up the clogged ones. We're trying to stop using emitters, as they're clogging and costing extra money.

Easy timer

Depending on the type of hydroponic system that you are constructing and where you are putting the plant growing device. You can need a single timer or two. If you use artificial lighting to make your plants grow the plants instead of natural sunlight, you'll want a timer to monitor the lighting system's on/off times. You'll need a timer to monitor the on/off times of the submersible water pump for flood and drain, drip, and aeroponic systems. Some types of aeroponic systems may require a special timer. Consult the aeroponic systems page to find out more about, and timers for, the forms of aeroponic systems.

Regular light timers work perfectly both for the lights and the submersible pumps. We do, however, suggest that the timer be rated for 15 amps instead of 10 amps. Fifteen amp timers are often referred to as heavy-duty; if not, check the back of the packet or timer for the 15 amp number. They typically have a cover and are generally water-resistant, so try to get one for outdoor use too.

We do not recommend digital timers, which are more costly than the analog dial type. Only because digital timers lose all memory and your settings if they lose power or get unplugged for some reason (unless you find one with a battery backup), these also have no more practical on/off settings than the same form of analog. Just make sure you have pins around the dial around the timer you get.

Air Pumps

Air pumps are optional in hydroponic systems other than in water culture systems. Yet their use has advantages, and the air pumps are relatively inexpensive. Anywhere they sell aquarium equipment, they can find air pumps. The air pumps provide the water and the roots with air and oxygen. Air is injected through air stones through an air pipe, producing a bunch of small bubbles that pop up from the nutrient solution.

The air pump helps to prevent the plant roots from suffocating in water culture systems as they are immersed 24/7 in the nutrient solution. The air pump is usually used within the reservoir for every other form of hydroponic device. This helps increase the levels of dissolved oxygen in the water to keep the water oxygenated.

Another advantage of using air pumps is that they keep the water and nutrients flowing and rotating as the air bubbles grow, this keeps the nutrients uniformly balanced all the time. The oxygenated water circulating also helps to reduce bacteria from gaining a foothold in the reservoir.

Grow Lights

Grow Lights are an optional hydroponic system feature. Depending on where you intend to position your hydroponic system, and where your plants can grow. You can either use natural sunlight or use artificial light to grow your plants. We recommend natural sunlight if you can make use of it, it is free and does not require any extra equipment. Though, if there's not enough natural sunlight to position your hydroponic device, or at that time of year, you'll need to use some artificial light to grow your plants at least.

Growing lights vary from those of most regular household lighting. Growing lights are designed to emit certain spectrums of colors that mimic natural sunlight. The plants undergo photosynthesis using these color spectrums (wavelengths) of light. Photosynthesis is needed for the plants to grow and produce fruit and flowers. So the form and the amount of light that a plant gets can significantly influence the ability of the plants to photosynthesize, and thus grow.

2.2 Making of Home-Based Hydroponic System

STEP 1

Nutrients make it via tubes through water push Assemble the hydroponic device. The device consists of six growing tubes made of 6 "PVC tubing, a stand, and trellis made of solid PVC, and a 50-gallon nutrient tank, a pump, and a funnel. The tank sits underneath the table of 6" PVC growing tubes, and the pump sits inside the tank to move nutrients up to the plants through a series of smaller PVC pipes. Every tube that grows has a drainpipe that leads back into the tank. The manifold then sits on top of the pipes and sends the tubes with pressurized water. Water is forced through a square of PVC, the funnel, to bring the nutrients to the plants in this device, and then fired out into tiny plastic tubes that run through each of the larger growing tubes. There are very tiny holes in the nutrient pipe, one hole between each plant spot. The nutrients remove the hole and spray the roots of the plant. At the same time, air bubbles are created by the flow of water so that plants get enough oxygen.

STEP 2

The tank contains approximately 50 gallons of water. Combine the nutrients and water in the tank Fill with water in the 50-gallon tank. Then gradually add two cups of nutrients to the tank (or as suggested by the fertilizer label), then turn on the required pump, and let the machine run for about 30 minutes to combine all the nutrients thoroughly.

Step 3

One of the easiest ways to plant a perfect hydroponic garden is to use purchased seedlings, particularly if you don't have time to grow the seeds yourself. The trick is to pick the healthiest plants you can find, and then remove all the soil from its roots. Submerge all the root ball in a bucket of slightly lukewarm water to cool water to wash the dirt off the roots. Water that is too hot or can be too cold will shock the plant. Separate the roots gently for the soil to come out. The little, tiny spray holes in the nutrient tubes may be obstructed by any dirt left on the roots.

Pull as many roots through the base of the planting cup after the roots are clean and then add expanded clay pebbles to keep the plant upright, and in place The expanded clay pebbles are strong, but they are also very light, so the plant roots are not harmed.

STEP 4

Bind the plants to the Trellis Using plant clips and string to bind the grille to the plants. The string will assist them in climbing straight up, which helps maximize the space in this confined area. Attach the string loosely to the top of the grille, add the clips and string to each plant's base, and gently wind the plant tips around the string.

STEP 5

Switch on the pump and monitor the daily machine Test the water levels daily; it may need to be tested twice daily in certain areas, depending on the water loss due to excessive heat and evaporation. The pH and nutrient levels are tested every few days. You don't need a timer because the pump runs full time, make sure the tank doesn't dry out, or the pump burns up.

STEP 6

Plants will cover the entire frame within a few weeks. Track Plant Growth A few weeks after planting, the plants will cover the trellis fully because they will have all the water and nutrients they need to grow rapidly. It is important to keep a close eye on the growth of plants and to tie or clip the stalks every few days.

Step 7

Check for signs of pests and diseases Search for Pests and Diseases Look for signs of pests and diseases, including the emergence of insect pests, chewed leaves, and foliar diseases. Because they are so similar to each other, one diseased plant will easily infect all the others. Immediately kill any diseased plants. Because hydroponically grown plants don't need to waste their energy trying to find food, they can spend more time growing. It allows them to be safer and stronger, and some of their strength can be used to fend off diseases. Since plant leaves never get wet when it rains, they're much less likely to get leaf rot, mildew, and mold.

Even though hydroponic plants are excellent at combating diseases, pests also need to be combated. Also, if it's hydroponic, insects and caterpillars can always find a way out into the greenhouse. Pick from and get rid of any bugs you encounter.

2.3 Advantages of Home-Based Hydroponic System

What if u were told that there was a way of growing plants faster, bigger, and using just 5 percent of the water usually needed.

Some people would believe this would be unlikely.

And yet hydroponics lets you do just that.

Hydroponics is a method for growing soilless plants, using only water, a nutrient solution, and a structure for keeping the plants up. While diverse forms of water culture have been practiced for several thousand years, the science behind hydroponics has been more thoroughly understood only in the last 100 years.

This has helped both domestic and commercial growers to grow plants in new ways with particular advantages and drawbacks.

This book will tell you all about the positive aspects of hydroponics that have undoubtedly contributed to the increasingly growing hydroponic cultivation industry.

Hydroponics forms part of a broader push to increase agricultural output, yield, and lower food production costs. Close behind this, indoor hydroponics grew, with an increasing number of enthusiasts increasing all kinds of plants at home.

Proper space management

Hydroponically growing plants need 20 percent less space than plants grown in soil. This means you can build more plants in a given area, or grow plants in tiny spaces where soil-based plants would not be practical.

This has significant consequences for the agricultural industry, where many plants are grown inexpensive indoor greenhouses, where effective use of space is key to obtaining a good return on the investment.

The main explanation for this is that hydroponic plants need less space than plants grown in the soil, and the roots do not need to expand in the ground to search for nutrients and water. Depending on the specific hydroponic process, water and nutrients are supplied directly to the roots, either intermittently or continuously. As a consequence, the roots are more compact and can grow closer together. Since less land is required, farmers, with fewer resources, may generate significantly higher yields.

No Soil Need for Hydroponics

The idea of growing soil-free produce was once a foreign concept but is now a reality for domestic and commercial production.

Growing soilless plants have a range of benefits. Soil quality varies greatly from one place to the next, and many plants have strong preferences for a specific type of soil. If you do not have this type of soil available, the importation of suitable soil or alteration of your existing soil can be costly and labor-intensive.

There are also a variety of areas across the globe that don't have access to the soil, or where land is limited. One of the first commercial hydroponic farming projects in the Pacific Ocean was on Wake Island. It is a rocky atoll which has no good soil for plant growth.

This island was used as a Pan American Airlines refueling stop during the 1930s. Importing fresh produce would have been prohibitively costly, so hydroponics was used effectively to develop the supplies required.

Other countries with little arable land, including desert or rocky areas, will no longer be restricted by how much they can expand. It is a driving factor for a hydroponic change, which is essentially why future farming is considered. In these areas, the production possibilities are significantly improved. This can reduce the need to import fresh produce, and can reduce water use, which in many countries can also be a problem.

Hydroponics Saves Water

Hydroponic Plants will grow with only 5-10% of the water required for soil growing. It is of immense benefit in areas where water supplies are limited and is a significant environmental advantage of hydroponic farming.

Hydroponics capitalizes on recirculated water, where plants consume what they need, collecting the run-off and adding it to the network. The only water that is lost is through leakage and evaporation, so if possible, a successful system will mitigate both.

Many hydroponic systems allow even better use of technology to reduce water waste further. The truth is, 95 percent of all the water plants take in from their roots is transpired into the rain.

As a result, some industrial hydroponic systems use condensers of water vapor to recapture this water and return it to the system.

Global food production keeps growing year after year and uses more water than ever before. We are endangering our planet's ecosystem unless we use technologies such as hydroponics to allow more sustainable agriculture.

Climate Control

Hydroponic environments offer absolute climate control. Temperature, light intensity, and period and even air composition can be modified, all according to what is required for optimum development. This provides an outlet for producing whatever the season, ensuring farmers can optimize production throughout the year, and customers can access goods whenever they wish.

Plants grow faster and bigger with hydroponics.

What is interesting about hydroponics is growth efficiency. You would think hydroponics would result in lower yields, but the opposite is true. There is room for faster growth than with soil, encouraged by the ability to regulate temperature, humidity, light, and nutrients.

Ideal conditions are produced to ensure plants obtain the right amount of nutrients that come into direct contact with roots. Therefore plants do not need to waste valuable energy seeking diluted nutrients in the soil. Alternatively, they should turn their focus to growing and producing fruit, leading to a better growth rate and larger plants.

Control over PH

Growers often ignore more Control Over pH levels, but it is a critical aspect of cultivation that ensures that your plants can access sufficient quantities of the nutrients they need to grow healthily.

Unlike growing plants in soil, the growing solution completely contains necessary minerals for growth. The pH of this solution can be easily balanced and reliably measured to maintain an ideal pH at all times.

Ensuring optimal pH increases the ability of a plant to pick up important minerals. If the PH levels change too much, plants may lose nutrient absorption capability. While some plants thrive in slightly acidic growth environments, the pH levels should normally range from 5.5 -7. Growers would be wise to investigate optimal PH levels for the plant in question and consider how hydroponic growth makes effective regulation possible.

No Weeds, Pests, or Disease

Any pests, plagues, or disease pests are time-consuming to eradicate from the soil and can have an effect on the growth of the plants you cultivate. There is no longer a concern with hydroponics. Soil-borne pests are likewise no concern.

Many hydroponic growing systems do not require pesticides as a consequence of the soil-free climate, which can make the product safer for human consumption and avoid the problems that pesticides can cause for the climate. Within a closed hydroponic culture system, you can take care of local variables more quickly.

Hydroponics Is Less Labor

While the setup costs of a hydroponic system are certainly costlier, whether, for domestic or commercial use, the labor involved in plant cultivation is greatly reduced. This frees up the energy to concentrate on other tasks, instead of tilling, hoeing, plowing, etc. The operating costs can also be reduced over time, but this depends on the system in question.

Climate is not a concern.

If you use a simple hydroponic system to grow a few tomatoes on your windowsill, or you operate a commercial hydroponic far away, you will remove a big cause of plant growth instability. Because most hydroponic plants are grown either indoors or in greenhouses, and all necessary water and nutrients are supplied manually, you remove the uncertainty that comes with unpredictable weather conditions.

Also, sunlight need not be a problem, as artificial rising lighting can substitute or complement sunlight. Using artificial growing lights will help you grow plants throughout the year. I use LED grow lights to help me grow tomatoes and salad greens during the whole year. The fresh salad is hard to beat whenever you need it.

Hydroponics Is a Good Hobby

A number of years ago, I developed an interest in hydroponics as a hobby, and I completely love it. Using hydroponics, you can start growing plants at very little upfront expense. You can use tons of DIY and pre-built systems, and this is very scalable.

You can start your windowsill by growing only one or two plants. That's how I began and helped me learn a lot about what plants need to grow and thrive. From there, you can scale things up, and there's no limit to how far you can take that hobby.

Though I do enjoy outdoor gardening, I love getting greenery in my house, and growing vegetables and salad greens during the year are so rewarding that I can use it to feed my family.

Urban Farming

Hydroponics is on the rise, with its global value valued at about $21.4 billion in 2015. Big global changes are on the horizon, which is expected to accelerate the growth of this form of farming.

Out of necessity, these changes will be introduced to accommodate a rapidly expanding global population. We are already using a large proportion of the land available to grow crops, so new farming techniques need to be created to increase yields or make another land suitable for crop production. Vertical urban farming is a prospective farming method that solves the problem of insufficient space and works with hydroponics very well.

Could hydroponics help?

While hydroponics receives considerable media coverage, many are left wondering whether it is a commercially viable activity. I guess the short answer to that question is yes, as demonstrated by today's large commercial operations.

2.4 Drawbacks of Home-Based Hydroponic System

TIME AND COMMITMENT

For success, hydroponic gardening requires some responsibility, diligence, and "stick-to-activity." Although this fantastic new approach reduces the labor needed to around 5-10 minutes a day, these are critical minutes. Your hydroponic garden will require some regular attention, aside from the occasional absence due to holidays.

Without destruction, an outer soil-based garden can be left to its own devices for weeks. It may be a mess of rotted fruit infested with weeds and bugs, but the plants can be revived in general.

A garden of hydroponics is not so forgiving. They usually need a little bit more TLC than that. You depend on your plants for their very survival. Yeah, you can be fantastic at automating a hydroponic garden, but you still have to oversee the entire operation and get away with the big trouble at the exit.

So, you have it in there. You are warned. One of the hydroponics drawbacks. Unless you are willing to make a fair effort to learn the method, interact with your plants, and make it all work, don't get into this. But if you're able to devote yourself to this noblest endeavor, the rewards are the frustration of the hydroponic disadvantages.

Okay, so I was "there, and I did it" By sharing my experiences with you on this website, I will save you a lot of frustration and time.

Still, I have to let you know right up front. Hydroponics is NOT precision science. I could tell you how to set up this thing step by step, list every piece of equipment and all the supplies you'll need, and post warning signs to look for. But No, You'll have to know by yourself, trial, and error.

I'm going, being honest with you. There were surprises with every hydroponic garden I'd ever planted. There are living beings that we deal with, and things don't always go the way they are expected. The main benefit of hydroponics is that you can play with a large pool of seed organisms. Heirloom tomatoes too! Many plants will blossom while others will fizzle. So it goes like this. This is the way gardening is, whether conventional or hydroponic.

I once planted a summer garden with a Bato bucket. I had so many beautiful cucumbers and peppers, I gave them away (and they were a tasty man). Yet my tomato seedlings. They were blooming like crazy; It was just so exciting. However, a single fruit never appeared. No real reason I can find out about that.

"Hydroponics is not organic – it's not even agriculture."

That was before organic regulations at the national level. At the time, organic agriculture was almost associated with local, small-scale farming, at least at home in New England. Helping one had to be supporting the other. The organic food industry has since become highly concentrated, with most of it owned by the same few companies that manufacture conventional food.

Organic agriculture grew exponentially but was disconnected from the effort to establish an alternative food system at the same time. The initial organic farmers and their backers changed gears when this occurred and started to stress the importance of location and scale. The local food movement was born in this way.

Sustainable agriculture as a movement is a reaction to the propensity of industrial agriculture to do away with the complex natural processes and connections that used to rely on farming.

The important distinction lies not in any specific set of techniques but the arrangement of the various systems. For example, in the past, plants and animals were raised at the same site, and fertility in the form of animal manure was restored to the soil. Today, in some parts of the world, animals are kept by the thousands in large buildings where their waste becomes an issue. In contrast, other sections of the world are turned into endless miles of grain, treated with synthetic fertilizer that pollutes the waterways. Countryside and country life in both cases vanish.

Sustainable agriculture – viewed as a method, rather than a collection of techniques – is intended to reorganize production in a way that preserves healthy agricultural environments and healthy communities. A significant part of this large project is organic farming practices. Consequently, it is disappointing that many customers, including my loyal client, regard organic farming simply as a strategy for producing food without pesticides.

This takes us to hydroponics – the method of growing soilless plants, either in a greenhouse or under artificial lighting, with the plant roots either suspended in water or protected by a neutral substrate like perlite or shredded coconut husks. All these nutrients are supplied directly to the plants in a form that can be absorbed immediately. Growing plants can be water- and energy-efficient in this way and the regulated climate ensures they can be made without herbicides or pesticides. The hydroponics scale can range from an apartment in Brooklyn to a business like Driscoll's that has hundreds of acres of hydroponic greenhouses.

The USDA currently requires an organic certification for hydroponic production, and they deferred deciding in late November as to whether they will continue to require it. Next National Organic Standards Board meeting is in April.

Many organic farmers are vehemently opposed to hydroponic organic certification. The apparent explanation for this is that organic farming activities focus on soil conservation, and hydroponics don't use any soil.

Greenhouses and containers are an important part of growing vegetables, and they are used as a stage by many organic farmers in their production systems. How does that mean if the goods are organic? It doesn't mean anything - the word doesn't even make sense when it comes to food. This makes sense in comparison to the entire systems of farming.

Hydroponics debate raises the major question about the word organic: is it intended to refer to sustainable agriculture practice or is it meant to refer to food products free of harmful chemicals? A lot of customers expect the latter to have work.

The bigger picture is that sustainable farming is basically about land stewardship. It is not food but farms, depending on how they manage the ground, that is, or is not organic.

Hydroponics is the ultimate separation of food production and nature, and the replacement of the last component for something that can be commodified and regulated without the need to take care of the natural environment.

Hydroponics may be a great way to grow food and maybe an integral part of how cities can feed themselves in the future. Still, it's no more a form of sustainable agriculture than growing wood fiber in a laboratory is a form of sustainable forest management.

Although the word organic has lost a lot of significance after the federal requirements kicked in, many proponents still equate it with a type of agriculture – one that as a method is intended to include a variety of land stewardship functions that are outside the scope of hydroponic development altogether.

In enabling hydroponics to be certified as organic, the USDA shows a very different view of sustainable agriculture from that of my devoted farmers 'market customers 20 years ago – and one that just as little overlaps with the concerns of organic farmers.

Bacteria Lack Effects

Plants that are grown in soil, rather than in hydroponic systems, have little exposure to numerous soil-based bacteria that coat plant roots. Plants were grown in the soil also produce immune factors to protect them from different forms of bacteria that are passed on to individuals who consume the plant as a consequence. Eating herbs, fruits, or vegetables grown in the soil over time will help to give you a stronger immune system.

Algae

One problem that arises in certain hydroponic systems on occasion is the growth of harmful algae in the water. In certain cases, the algae will bloom and die so rapidly that it will accumulate and suffocate on plant root systems, rendering the plants vulnerable to pathogens.

Price

It has been suggested that hydroponically grown crops and plants are likely to be costlier than conventionally grown counterparts. This is due in generally large part to the wide range of equipment and advanced materials needed to sustain hydroponic systems.

Myths about Hydroponics

Research's Adverse Effects has been carried out measuring the quality of hydroponically grown products against traditionally grown products. There was no difference between the two in taste, visual quality, or texture, according to a University of Nevada experiment. Hydroponic growing systems, in addition, remove many of the problems associated with conventional growing methods, including plant exposure to harmful pests and soil discord.

Hydroponic manufacturing is limited.

Although you can grow all year round (and this the make up the difference), you are limited by the available space. When overcrowded, a hydroponic plant can't survive. Where plants can be grown right next to each other in more conventional growing practices, a hydroponic plant requires space to spread out. It ensures that there is a far smaller number of crops that can be grown at one time than those produced in a field.

Daily control

Nearly daily control of a hydroponic greenhouse. Holding a hydroponic garden's delicate balance in check may be overwhelming to some, and most hydroponic farmers will have unsuccessful crops in their first few attempts before they have their methods under control. Be able to learn from trial and error. This is not an easy job to undertake, but ultimately the benefits are fantastic.

Water filter n cycle

As the water is filtered and cycled in the entire hydroponic system, it can kill a whole crop in a matter of hours if only one disease takes hold. If you find that your hydroponics crop contains a diseased plant, the chances are that the rest of your crop is not healthy either. Many hydroponic farmers thus lose whole crops, making disease control an ever-important part of the farming routine for hydroponics.

Water-based microorganism

The danger of water-based microorganisms is another challenge to hydroponics farming. Fungi and bacteria grow in water. Although some bacteria and fungi are perfect for crops, others may be hazardous. Unfortunately, warding one off without the other is almost impossible. The sterility of farming with hydroponics is just as strong as the farmer's sterilization process. Holding your crops free of harmful microorganisms and healthy eating can in itself be a full-time task.

Power outages

Hydroponic farms face power outages. Since most of them have water and light set on timers, if the power goes out, and remains out longer than the backup generator will operate, then the whole crop is at risk of destruction. Most farmers in hydroponics tend to have many long-lasting backup generators on hand, just in case of an emergency.

Expensive farming

Hydroponic farms are not inexpensive to create. The equipment needed to operate the regular operation of timed watering and illumination, the filtration system, and tanks involved, and they can cost tens of thousands of dollars-even hundreds at times! If you are actually serious about starting a commercial hydroponic plant, you will also find that in some cases, they are difficult to insure, as there is still so much uncertainty about the danger involved in growing and running these farms.

Because you can now see, there are many great benefits of hydroponic cultivation, and others in terms of drawbacks. If you still feel that hydroponic agriculture is something you might be interested in, many fantastic online tools can help you get started. You may also contact your local agriculture department to find out what your local farm operating conditions are, as well as any local threats for common diseases, fungi, or bacteria.

2.5 Best practices of Home-Based Hydroponic System

To live, humans need air, food, water, and living space. Such issues are not infinite, and thus humans are dependent on land area optimization and biodiversity protection. The human population is growing, and it is expected to increase from 7.0 billion to 9.5 billion people in the next 40 years (Sahara Forest Project, 2009). It means an ever-increasing demand for food products, and it is projected that food production would have to be increased to supplement and provide accessibility for everyone. The term "Hydroponic" describes any means by which plants can be grown through a process that does not require soil use but includes inorganic nutrients or nutrient solutions. Gericke, who defined plant cultivation methods in liquid media (nutrient solution), coined the term Hydroponics. In addition to Gericke, during the thirties, several attempts were made to follow the methods of soilless growing plants. However, due to insufficient knowledge of the nutrients and high costs involved in the process, technical advancement was too inadequate. Despite all this, countries like the USA and others were keen to implement this technology so that it would be possible to grow plants indoors without the favorable soil needed as well as to con-troll the nutrient. One of the basic principles for the growth of vegetables, both in soil and in hydroponic systems, is to provide all the nutrients the plant requires.

In seventeen elements, various chemical elements are important for plant growth and production: carbon, hydrogen, oxygen, nitrogen, phosphorus, potassium, sulfur, calcium, magnesium, manganese, iron, zinc, boron, copper, molybdenum and chlorine. In hydroponic crops, absorption is typically proportional to the concentration of nutrients in the solution close to the roots, being greatly affected by environmental factors such as salinity, oxygenation, nutrient solution temperature, pH and conductivity, light strength, photoperiod and air humidity (Furlani et al., 1999). In addition to these details, the advance of Hydroponics is commercially encouraged in India. Letcreta Agritech is Goa's first start-up indoor Hydroponics that produces 1,5-2 tons of high quality, pest-free, leafy vegetables. Bit Mantis Innovation offers a GREEENSAGE IoT solution, i.e., a micro edition package using a hydroponics system for water use and nutrient that is ideal for the convenience of the user. Junga Fresh n Green, a 9.3-hectare start-up for hydroponics farming in Shimla district with a joint venture WPC, Netherlands, is heading towards safe greenhouse environments.

This paper aims to show the hydroponics specifics that are applied using electronic circuits, water, and nutrient solution, i.e., soilless solution. The machine supplies nutrients automatically, and nutrients can be tracked. Corrections can be made after the testing is completed, i.e., it can be monitored accordingly, resulting in higher efficiency. To make the automated model more versatile, we plan to incorporate pest detection and link to Wi-Fi (IoT-based) node. This system saves water and fertilizers, performs more than the soil system in contrast.

Indoor Plant Growth

It can be a great challenge for making some plants and vegetables in remote areas such as deserts of north and south pole. Very few plant species survive in these circumstances and are therefore not used as a source of food. We've developed a system in this study that can grow common plants and vegetables and work without relying on the outside environment. We accomplished this by using the Hydroponics technique. Hydroponics is a plant growing process without the use of soil. The device was automated with the assistance of microcontrollers and sensors to reduce human interference. An Internet of Things (IoT) network was developed to enhance reliability and allow remote monitoring and control where appropriate. The consumer need only plant a seedling and set the initial parameters. When completed, the device will maintain the settings and encourage the healthy growth of the plants.

The goal of the Titan Smartponics system was to build a fully automated hydroponic system, which was low cost and relatively easy to operate for the average consumer. This task was achieved by the use of Arduinos, a Raspberry Pi, open-source software, and a few sensors. The automated hydroponic device retained the conditions required for the test plant to survive and was able to integrate remote monitoring and control IoT networks. A few benefits of the Titan Smartponics system is that there is full control over the aspects that allow a plant to grow, it can be modified to meet the needs of a variety of plants, and it doesn't depend on the outside atmosphere or climate to thrive. Titan Smartponics proved its value relative to other systems through its full automation feature and its ability to be small enough and cheap enough for consumer use.

Choosing What to Grow

It can be grown hydroponically on just about any plant, but for beginners, it's a good idea to start tiny. Herbs and vegetables that grow quickly need little maintenance and don't need a wide range of nutrients are the best choices. Fast-growing plants are best because they make it easy to determine how well your system is working and change it as needed, and Waiting months for harvest time can be a major letdown to find out that your machine is not functioning properly. Maintenance-free plants are perfect for beginners as they allow you to focus on learning about your system — you can then move on to more complex vegetables. When you are growing a variety of plants, it is also important to make sure that the nutrient requirements are compatible so that they grow well together.

Lighting

Hydroponic systems are mostly installed indoors in areas where direct sunlight is not available during the day. Many edible plants need at least six hours of sunlight each day; even better, 12 to 16 hours. If you have a sunroom or other space with lots of window exposure, supplemental rising lights are likely to be needed. Hydroponic device kits normally come with the requisite light fixtures, but if you're putting your components together, you'll need to purchase separate light fixtures.

HID (High-Intensity Discharge)

light fixtures are the perfect lighting for a hydroponics device, which may include either HPS (High-Pressure Sodium) or MH (Metal Halide) bulbs. The light from the HPS bulbs emits a more orange-red color, perfect for plants in the stage of vegetative growth.

T5 is another form of illumination used in hydroponic rooms. This produces fluorescent, high-output light with low heat and low energy consumption—ideal for growing cuttings of plants and plants with short growth cycles.

Put the lighting device on a timer so that the lights turn on and off every day at the same time.

Room conditions

The setting up of a hydroponic device under the right conditions is very critical. Key elements include relative humidity, temperature, CO2, and airflow. The perfect humidity for a hydroponic growing room is relative humidity from 40 to 60 percent. Higher levels of humidity - particularly in rooms with poor circulation of air - can lead to powdery mildew and other fungal problems.

Ideal temperatures range from 68 to 70 degrees Celsius. High temperatures may cause the plants to stunt, and if the temperature of the water is too high, root rot can occur.

There should also be ample supply of carbon dioxide (CO2) in your grow space. An excellent way to ensure this is by maintaining a consistent airflow in the room. More advanced hydroponic gardeners should supplement in-room CO2 rates, because the more CO2 you have available, the faster your plants will grow.

Water quality

The ability of water to provide dissolved nutrients to your plants can be influenced by two factors: the number of mineral salts in water as determined by PPM, and the pH of water. Rough water with high mineral content does not absorb minerals as easily as water with lower mineral content, so if it is high in mineral content, you may need to filter your water. In a hydroponic device, the optimal pH level for water used is between 5.8 and 6.2 (slightly acid). If your water fails to reach this standard, chemical substances may be used to change the pH to the optimal range.

Nutrients

The nutrients (or fertilizers) are then used in hydroponic systems in both liquid and dry forms, as well as organic and synthetic forms, are available. The substance you are using will contain all the main macronutrients – nitrogen, potassium, phosphorus, calcium, and magnesium – as well as the important micronutrients that include trace quantities of iron, manganese, boron, zinc, copper, molybdenum, and chlorine.

Using fertilizers designed for hydroponic gardening, if you use them according to the directions of the box, you should get good results. In a hydroponic method, stop using traditional garden fertilizers as their formulations are formulated for use in garden soil.

Pick hydroponic nutrient items that are customized to your unique needs. For example, some are marketed as best suited for flowering plants, while others, such as leafy greens, are better suited for promoting vegetative growth.

Optional Equipment

Beginners should try to invest in a few additional items in addition to the standard hydroponic system.

To check the water's PPM and pH, as well as the room temperature and relative humidity, you will need meters. Some combination meters are available, which will measure the temperature of pH, PPM, and water. You can also buy meters that measure the temperature and moisture in your growing room.

Depending on your environment, you can need a moisturizer or dehumidifier to change the relative humidity in the can room optimally.

You may also want some form of fan or air circulation equipment to boost your grow room's airflow. Only a simple oscillating fan does work well, but you may want to invest in a more advanced intake-and-exhaust system as you get more experience.

Good Starter Plants

Some plants that generally work very well for beginners who are still learning the fundamentals of hydroponic gardening have greens such as lettuce, and spinach, Swiss chard, and more of kale herbs such as basil, and parsley, oregano, cilantro, and mint Tomatoes Strawberries Hot Peppers Systems For More

Experienced Gardeners Two more complex systems are best reserved for hydroponic gardeners who already have these systems.

NFT

Device NFT stands for Technique of Nutrient Film. This requires a solution of water and nutrients that continuously flows through a rising tray in a loop from a reservoir, where all plant roots are suspended in thin air and absorb nutrients as the solution flows by. If something is wrong with the pump mechanism, when the flow stops, the roots will dry easily, so this device needs a user who is able to control the machinery and repair it easily if problems occur.

Aeroponic system

An aeroponic system is a high-tech process by which plant roots are suspended in air and misted with a solution of water and nutrients every few minutes. It is a system that is highly efficient but includes sophisticated pumps and misters. If there are issues with the machinery, the plant roots can dry out and die quickly.

Chapter 3: Factors Affecting a Hydroponic Garden

Most crops are grown in outdoor settings with sufficient sunlight where native soil and appropriate irrigation and fertilization practices can provide for plant needs. Soil plays many important roles for plants, including providing physical support, providing water holding power, supplying plants with many of their nutrient needs, and supporting the biological activity required for nutrient cycling. Soilless cultivation is a plant-growing system that provides many of the same functions as soil by physically supporting the plant and providing a rooting atmosphere that gives access to optimum water and nutrient levels.

Soilless processing can take place through industrial processes (perlite, vermiculite, Rockwool) in naturally occurring sand, peat moss, coconut husks (coir), and materials produced from rocks or minerals. Products available locally in many areas, such as composted pine bark, rice hulls, and other materials are used in soilless culture. Some soilless production occurs in foam substrates as well. The following sections will discuss these products in greater detail.

Production of soilless can be carried out in many ways. Often plants are grown in a substratum (simply the material in which the plant roots live) that mimics the physical support and water and nutrient supplying roles of natural soil but is not soil. An example would be poinsettias, bedding plants, vegetable transplants, and other crops grown on a peat-based substratum in a greenhouse. These plants are fed, fertilized, and managed to maximize growth by maintaining the substrate's physical properties and supplying all the plants with water and nutrients.

Soilless development can also include roots not produced in a substratum. Nutrients are dissolved in water in these solvent-based processes, and plant roots are bathed directly in a nutrient solution.

The authors find that the words "soilless growth" and "hydroponics" should be used interchangeably for the purposes of this publication.

These terms are used to describe plant production systems where natural soil is not used for the An Introduction to Small-Scale Soilless and Hydroponic Vegetable Production. There are two subcategories in this publication within soilless development or hydroponics. The two subgroups are 1. Systems to cultivate the crop using a soilless medium or substratum, and 2. Systems that use only a nutrient solution to grow the crop without any medium or substrate other than to expand the plug for transplants. Items sold in stores as "hydroponic" could be using either a production device form of soilless media or nutrient solution. In no conditions should a crop be considered hydroponically grown when grown in natural soil?

3.1 Design of your system

Each system is a little different, and it took more than 14 years of experience in hydroponic gardening to shape these opinions. Understanding the growing hydroponic systems problems should help you understand the details below better.

To grow healthy plants-water, nutrients, and oxygen-three items are required at the root level. The growing section includes a link to learn exactly how each device works to provide these items to your plants.

DWC and NFT Homemade

Hydroponics Design Such a system design may be DWC, NFT, aeroponics, drip method, or Flood and drain depending on how the device is used Deep Water Cultivation Systems (DWC) and Nutrient Film Technique Systems (NFT) use low-pressure pumps, and are thus less likely to leak. The DWC / NFT concept is more economical to build than others. These systems are reliable and low-maintenance, must be tested once a day only. Both systems, such as expanded clay pellets in netted pots, can be designed to use a reusable grow medium. These homemade hydroponic systems will produce outstanding results. Discover more about DWC and NFT programs.

Flood and drain systems

It is Also known as the Ebb and Flow system; this design fills the table with water for 1 to 20 minutes to fully soak the media (repeated as needed) Flood and drains systems often use low-pressure pumps, making this another inexpensive homemade hydroponic concept to be installed. When roots or debris get into the drain, it may lead to a situation of flooding. This question can be solved mainly by placing the drain from the closest planet in the system at least 12 inches apart. Such devices are usually low-maintenance and need to be tested once or twice a day.

These homemade hydroponic systems, such as expanded clay pellets in netted pots, may be designed to use a reusable grow medium. Clay pellets must be submerged every two hours for 20-30 minutes to stay sufficiently wet. This includes monitoring the hydroponic pump using a timer. This design is only slightly less robust than DWC or NFT systems, due to the timer (which may fail) and the risk of clogged drains. Flood and drain systems can yield excellent results, too. Know more about the drainage and flood schemes.

Drip / Spray Systems

The use of drip heads, spray nozzles and high-pressure pumps make these systems more expensive to install than other homemade hydroponic systems with a drip feeder line (or two) on each plant; they can easily be operated as a drip system (using PVC pipes for water return). Having a high-pressure pump also makes it easier to leak out further. Drip/spray heads that travel beyond their set positions can also cause leakage. Drip/spray heads are infamous for clogging, making maintenance of the system more a concern. These systems need to be tested 2-3 times a day to avoid leakage and to ensure that nutrient solution is obtained for every plant.

Such systems, including expanded clay pellets in nested containers, can be built to use a reusable grow medium. Because of the risks of leakage and clogging, these homemade hydroponic systems are less reliable than DWC and NFT systems, unless the extra care/maintenance is required. These systems can yield excellent results when nozzle clogs are detected and fixed in a timely fashion. Know more about the devices that leak.

Aeroponic Systems

The plants are grown in an aeroponics device in just AIR! The roots must be regularly or constantly bathed with nutrient solution. Aeroponic systems are spray systems that grow plant roots in an empty air space without any other growing medium. High-pressure pumps are used by aeroponic systems, which make the device more likely to leak. The use of spray heads and a high-pressure pump makes the construction of this type more costly. Moreover, at some point, spray heads are likely to clog up.

With no growing medium to protect the plant roots, in less than an hour, a clogged spray head will lead to a dead plant. Pump failure or power outage will cause all of your plants to die within one hour. It makes high maintenance construction of the aeroponic system- these homemade hydroponic systems should be tested four times a day or more. These systems are capable of providing excellent performance, despite the demanding amount of attention and maintenance. Read more on aeroponic devices.

3.2 Weather Conditions

You can build the ideal "micro-climate" for your plants using a hydroponics device. The benefit of this cycle is that you can reap a reliably bountiful harvest, season after season, in providing the perfect food and environment.

Now a spare bedroom isn't ideal for growing vegetables automatically. To be effective, you need to provide sufficient light, fresh air, and a proper temperature and humidity range to your Hydro system.

Equally significant is the time spent watching, relaxing, and providing some TLC in the space!

How to Hydroponics:

Fresh air

If your garden is in a secluded bedroom or grow small space; keep the door open, and the window cracked. Air trade is an absolute must. Plants deplete the CO2 (carbon dioxide) in the atmosphere to provide fresh air, one way or another—a must-see fan.

Light

If your garden is sitting in direct sunlight for a few hours per day, you'll need to install additional lighting above the growing area. Indoor increasing lights are now so advanced in the spectrum they offer that they are practically similar to sunlight. You could grow vegetables in a dark cellar! Learn here about these beautiful lights.

Most of all, have indoor houseplants successfully developed without any special lighting. House lamps and window lighting for Schefflera or spider plants are plentiful, but it is not enough to render tomatoes and squash.

You'll need to have good lighting for your hydroponics food factory. The development of vegetables needs very bright light and contains the appropriate form (spectrum) of light. Otherwise, they are not going to flower and fruit. Sunlight, special HID lights (high-intensity discharge), or latest generation multiband LEDs can do this.

Vegetable plants in an outdoor garden need 4-6 hours of direct sunlight, with 8-10 more hours of bright indirect natural light every day. Indoors, with special (and relatively expensive) HID or LED lamps set on timers to run 14-16 hours a day, you can almost replicate this.

As with the weather, when it comes to lighting up your backyard, more isn't better. You would think it would promote better or faster growth if the lights were held on 20 or even 24 hours a day. Op!

Plants make use of the light and heat throughout the day to generate their energy. At night, they assimilate and expand when the lights go off, and they cool down. Every night, plants need to rest and heal just as you do. Otherwise, they get tired and die quickly.

Temp & Humidity

For your plants to survive, you must also have the appropriate temperature range and humidity. This is a very simple necessity often ignored by beginners.

Only designating a spacious and well-lit area for your new hydroponics garden isn't enough. For optimum output, you must have the right microclimate.

Proper humidity and temperature also repel rodents, bacteria, fungus, and disease.

Okay, then maybe by now you're sick of hearing this. But that's just so important! The proper temperature range for a first hydroponic garden is such an essential and crucial factor that we don't want you to miss it! Keep in mind the optimal magic range: 65- °.

There are two types of vegetable crops, and they require a different temperature range:

Tropical seasonal crops and herbs: [tomatoes, peppers, eggplants, cukes, beans, squash, melons, herbs] Daytime: 70- degrees; Nighttime: 60- degrees.

Pleasant seasonal crops: [broccoli, cabbage, lettuce, endive, peas, spinach, green onions]

Daytime: 60- degrees; nighttime: 50- degrees. Absolute minimum temp=40 degrees Yes, by sacrificing on temperature, you can mix them up some (like growing lettuce with tomatoes). But in general, if you grow seasonal crops together, grouped as above, you'll get the best results. And you have the best temperature range to live inside.

3.3 Water Supply check

Water quality in a hydroponic garden is one of the deciding factors for the result. In hydroponic systems, chronic issues are often traced back to the water supply. Water is the essential transport mechanism in a hydroponic garden because it dissolves and carries nutrients to plants and their root systems. Hence, it is necessary to consider the consistency of the water supply. Water also dissolves impurities and not just nutrients. These impurities cannot easily be visually detected and can be harmful to plants. Hydroponic growers usually make incorrect assumptions about water purity, based on a sample's clarity. Assuming the crystal-, odorless water they get from the tap is pure water, and their plants will be healthy in turn. After all, drinking is healthy for humans. so why not your plants, right?? Incorrect. Plants appear to be more prone to such treatments with water than humans. In "Good Drinking Water," impurities are often left, which are a problem for your plants, particularly in a hydroponic system. Town water supply requirements vary from city to city for drinking water. City water sources are usually formulated with additives to make them drinkable. Such specs of drinking water refer to make water safe for humans but not necessarily safe for plants.

The EPA has a fixed limit of up to 500 ppm allowed in the water, and it is still considered safe for human use. Calcium and magnesium (or the water's "hardness") make up a majority of the dissolved solids in the water, so how much of the ppm is calcium or magnesium, and how many other contaminants are there? There are many other contaminants inside your water that can be sources of ppm, including agricultural and urban runoff, industrial waste, sewage, and natural sources such as leaves, silt, plankton, and rocks. It has also been established that piping and plumbing leak metals into the water, which also contributes to the ppm. You'll check the water, for example, and the ppm is 500. The water tastes excellent for you and others, and you think watering your garden with it would be fine. Now, you're in an early vegetative stage and want to keep your formula around 700 ppm for feed. The issue is that your water is already at 500ppm, and you add 200ppm of good nutrient content, while 500ppm (over double) is unknown solids that are fed to your plant!! And this is what the problem is! You won't be able to monitor the number of nutrients your plants get precisely. Starting with pure water, as near as you can get to 0 ppm, it allows you to add any part of the nutrient formula in the exact amounts needed during and growth process. Being able to monitor the quality and ppm of your plant food accurately will give you the potential to achieve impressive results and have good harvests each time. Starting with pure water can also help you prevent issues with nutrient shortages or lockouts. When you're not starting with pure water, we recommend getting your water checked in your garden before using it.

1. Well, Pond, Lake, and Stream Water & Pathogenic Contamination

Many sources or forms of water are available: tap water, reverse osmosis (RO), bottled water, rainwater, well water, lake/pond, and even the stream water. Every form or source of water shall have an advantage or disadvantage. The type of problems that growers will face will usually depend on their water supply. Inadequate water quality can lead to various issues such as mineral toxicity/deficiency, stunted growth, build- of mineral/salt, pollution of bacteria, etc. The good news is that water quality problems, in most cases, have clear solutions which do not require complicated procedures or techniques. Also, beginner growers can use a few smooth but successful methods to solve their water quality problems correctly. Many factors can influence the quality of the water, familiar sources and considerations include 1. Well, Pond, Lake, and Stream Water & Pathogenic Pollution Well, pond, lake, and stream water appear to have in them bacteria and soil- pathogens (i.e., Pythium, Fusarium Wilt, etc.) harmful to your plants. Anaerobic bacteria and pathogens can give your plants several diseases, which can be challenging to treat if not impossible. Chlorination is said to be probably the most common form of bacterial and pathogenic contamination treatment. However, you will need to allow the chlorine to dissipate in your garden before use. Hydrogen peroxide is also useful as a sterilization process, but must also be consumed before use. Hydrogen peroxide can kill your plants in significant quantities. However, in smaller amounts of roughly 5ml per gallon of water, due to the extra dissolved oxygen in the water and nutrient solution, hydrogen peroxide can be beneficial for plants. Microbes that are healthy, or aerobic, can also help kill the unwanted pathogens when they feed on them.

2. Rainwater

Rain can be a free, clean water source, depending on how it is collected and stored, and where. Rainwater can also contain pollutants if it is not collected or stored in the appropriate receptacle. Areas with high emissions can create acid rain as a result of raindrops absorbing and storing contaminants from the clouds on their way down. Rainwater gathered from a galvanized iron roof, or rain gutter may contain high zinc levels. Water stored in new cement tanks may contain minerals leaching from the cement into it. We advise you to check your rainwater before you use it in your garden.

3. Hard Water and Excessive Minerals

Many water bodies are seen as hard water. Water is considered 'strong' when it contains large quantities of calcium, magnesium, and/or other dissolved elements. While plants need calcium and magnesium, water that contains too much (referred to as "Full Hardness") may cause serious problems. Tell the local water supplier to get an overview of your water supply. When you are using well water, when you give them a sample, several laboratories will provide you with an analysis. When you have 200 ppm or more of dissolved solids in your water supply, we highly suggest that you get a water analysis to assess the substance. Excessive calcium is the primary determinant of whether or not water is rough. If an investigation shows that your water supply's calcium content is greater than 70 ppm, you can consider using a nutrient specially formulated for hard water (i.e., FloraMicro hard water). Nutrients expressed for hard water provide plants with a combination of chelated micronutrients composed explicitly for the conditions of hard water.

Hard water can cause physical issues, as well. Among other components, high levels of iron, calcium, and limescale start to scale up on water pumps, tubing, heating elements, reservoirs, drip emitters, etc. It causes a lot of problems with clogging devices, ceasing to function, and can also cause pH issues. Water softeners can be used to treat rough water. Water softeners use an ionizing resin that flushes back by sodium chloride (salt), substituting sodium ions for calcium and magnesium. This method would also add a small amount of sodium to the water, which is poisonous to plants, so it is advised to use potassium chloride in your softener as opposed to sodium chloride, which is more widely used. A reverse osmosis (RO) system should be able to eliminate pollutants without adding anything to the water, but most home-RO systems only produce one to two gallons per hour. You would usually want to use a RO system to plant, which can supply 100- gallons per day. RO systems can provide a water quality similar to that of the distilled water, depending on the RO system and proper filter adjustments. Distilled water is pure water, free from bacteria and viruses, but free from minerals or trace elements, too. When you use distilled water, you'll need to apply around 200- ppm of calcium and magnesium to make up for that. RO water needs this, too. Starting with clean, pure water can give a safe start to your plants. Mature plants, which are fed with water from an unpure source and encounter problems, undergo significant health changes, and development when transferred to pure water.

4. Chlorination

City and municipal water treatment facilities more commonly use chlorination to monitor the bacteria and contaminants rates. If the water contains high levels of active chlorine, it can cause damage to several crops, particularly raw fruits and all vegetables such as those salad greens, strawberries, and others. The very good news is chlorine is highly volatile, and when it comes into contact with air, it evaporates quickly. This can be done by putting chlorinated water in an open aerated tank and allowing the chlorine to dissipate 48 to 72 hours (depending on chlorine concentration) before mixing the nutrient solution with it. When active chlorine levels fall below one ppm, the plants are safe to use. A water softener may be used to extract chlorine from the water, but bear in mind the issues that come with a water softener (see 'Hard Water' above). You'll also want to make sure the water does contain chlorine, not chloramines. Since before the 1950s, many municipal water treatment facilities have been using chloramines, and in 2000 there was also a federal order to convert the largest U.S. cities to chloramines. Through distillation, you cannot get rid of chloramines, let the water sit, or even have a regular RO method. With special RO systems designed for filtering chloramines (i.e., a 5-stage RO- system), chloramines can be extracted. Chloramines may come in more regular contact with carbon and have to be separated for long periods of time. RO systems built for chloramines typically have more than one carbon filter, allowing water to pass slower than a normal RO system through the filters.

7. Iron and Iron bacteria

Iron in the form of iron hydroxide typically occurs in groundwater sources close to areas of iron sand or iron ore deposits. Iron hydroxide is not specifically harmful to plants but can cause problems with blockages in various components of your system. If iron bacteria are not extracted, a variety of nutrients will be consumed, which are supplied in watering systems for plant production. Iron hydroxide may be extracted by aeration and settling, or by flocculation with various chemicals. Iron bacteria may be extracted by water- or nutrient solution sterilization.

8. Water temperature

The nutrient solution will have a temperature of 68 to 72 degrees Fahrenheit. It is a very good idea to allow it to get to the same temperature as the water in the reservoir before adding water to your reservoir. Plants do not like the rapid changes in temperature, especially in the root zone. Aquarium heaters can be used in the winter to warm the nutrient solution and to search for "chillers" to cool down the solution in the hot summer when high temperatures become a concern

3.4 Check for Substrates

Growing soilless plants, the accepted concept of hydroponics, continues to gain popularity in commercial horticulture, more and more products are being created for it. In the last five years, more advanced lighting, simpler mixing of nutrients, and streamlined plant supports have all come onto the market. But the advancement and popularization of an alternative that media has become one of the most exciting developments in the hydroponics world.

Possibly there are hundreds of different types of growing media; generally, anything a plant may grow in is called a growing medium. Rockwool / stone wool (the industry standard), oasis blocks, vermiculite, perlite, coconut fiber (coir), peat, composted bark, pea gravel, sand, and expanded clay, lava rock, fiberglass insulation, Sawdust, pumice, foam chips, polyurethane rising slabs, and rice hulls are amongst the aggregates now available. -- Alternative has positive and negative characteristics, and the choice between aggregates depends on several factors, including the size and variety of plants you wish to grow and the variety of hydroponic systems used.

Rockwool / Stonewool Industry Quality.

Made from those rocks that have been melted and spun into fibrous cubes and rising slabs, Rockwool has an isolating structure and provides a good balance of water and oxygen to the roots. Rockwool can be used for continuous drip or ebb and flow systems, and is suitable for all sizes of plants, from seeds and cuttings to large plants.

Rockwool is considered the perfect substrate for hydroponic growth by many commercial growers. Due to its unique structure, Rockwool can hold water and retain enough air space (at least 18 percent) to promote optimal root growth. Because Rockwool exhibits a slow, steady drainage profile, the crop can be managed more specifically between vegetative and generative growth without fear of drastic EC or pH changes. Note that certain Rockwool products need overnight water soak before use, as the bonding agents used to shape slabs will result in high pH.

Vermiculite / Perlite

Perlite is a material that is made of volcanic rock. This is pure, lightweight, and sometimes used as a soil nutrient to improve aeration and soil drainage. Vermiculite, which is used in the same way as perlite and is frequently mixed, is made from mica expanded by heat and has a glossy, flaky look. Since perlite and vermiculite are so lightweight, only seeds and cuttings are suggested for launch.

Perlite has good wicking action, which then makes it a good choice for hydroponic wick-type systems, plus it is relatively cheap. The greatest downside to perlite is that it doesn't very well hold water, meaning it can dry out easily between waterings. The exact opposite is true of vermiculite; if used straight, it retains too much water and can suffocate the roots of the plant. In addition, perlite dust is bad for your safety, so wear a dust mask while handling this medium.

Media Alternatives

The increasing cost and difficult disposal of Rockwool have led many farmers to consider alternative substrates. With so many choices available, there is virtually one substratum for each situation. The following are only a few of the more common and promising choices.

Expanded pellets made of clay

This human-made product is also called growing rocks and acts as a growing medium extremely well. It's made in a kiln by baking clay. The scale of the pebbles varies from 1 to 18 mm and is inert.

Clay pellets are lined with tiny air pockets, giving them decent drainage. Clay pellets are ideally suited for ebb and flow systems or other regular watering systems. Due to the lack of good water-holding ability of expanded clay pellets, salt accumulation and drying out may be common problems in poorly operating systems. It is recommended that clay be flushed periodically with either a half-strength nutrient solution or a commercially available flushing agent.

While pellets are very costly, they are among the few media types that can be easily reused. Remove old roots after harvest, and sterilize with bleach, steam, fire, or hydrogen peroxide.

Land

Land and Sand, the oldest known hydroponic substrates, is not commonly used today, mainly because of its poor ability and weight to carry water. Sand appears to stack tightly together, decreasing the amount of air available to the roots; therefore, it is ideally suited for hydroponic use by a coarse builders 'sand. Conversely, for improved water holding capacity and lighter weight, and can be combined with other material.

Gravel

Gravel was one of the first hydroponic systems used in commerce. Gravel is usually fairly low-cost, works well, and is usually easy to find. Gravel provides the roots with plenty of air but does not hold water, meaning that roots can dry out easily. The weight makes handling challenging, but it has the advantage of not breaking down in shape and can be reused.

Gravel can be quickly reused when being washed and sterilized between crops. For cleaning can using flame, steam, bleach, or hydrogen peroxide.

Sawdust

As a hydroponic medium, Sawdust has had limited success, but it is used with tomatoes, particularly in Australia. Many variables determine how well Sawdust works, mostly the type of wood used and its purity. Growers need to be careful not to contaminate their Sawdust from wood processing facilities or unwanted tree species with soil and pests or chemicals. Another sawdust problem is it'll decompose. Sawdust still holds much moisture, so be careful not to overwater. The best thing about Sawdust is that it's safe.

Coconut thread

Coconut fiber, also known as coir, is quickly becoming one of the world's most important rising media and could soon be the most important. It is the first totally "real" medium in hydroponic systems, which offers top efficiency. Coconut fiber is a coconut industry waste product, which is simply the pulverized coconut husks. Coconut fiber is available in various grades, with the lowest having an extremely high salt content which needs leaching before use.

The coconut fiber's key advantages are its oxygen and water-holding capacity. It can retain a greater capacity for oxygen than Rockwool, but it also has a superior capacity for retaining water. Some research has also shown that coir can possess abilities to repel insects. High-quality coir (the degree widely used for hydroponics consists of the coarser fibers) often has the advantage of not having any, or very low, nutrient levels, so it does not alter the nutrient solution composition.

Cubic oasis

The oasis rooting cubes are rigid, open-celled, water-absorbing foam parts engineered specifically for optimal callus and rapid root formation. As it is made from phenolic foam, oasis cubes are most commonly used in commercial floriculture as rooting media and are a great medium for starting seeds and cuttings in hydroponic production. Oasis cubes bear their weight in water more than 40 times and have wicking motion that pulls water to the foam's edges. They have a neutral pH and can easily be transplanted into almost any form of a hydroponic system or through media.

Place mouse Sphagnum

A fully natural medium that is used in most soilless mixes as the main ingredient, sphagnum moss, is frequently ignored as a hydroponic medium; however, it has many properties that are highly suitable for hydroponic growth and is readily available.

Sphagnum moss has long strands of highly absorbent, sponge-like material that hold and retain large amounts of water while providing good aeration at the same time. Because of this structure, it is best used in the development of bigger lattice or net-pot where the long strands will spill the holes in the pots to wick up water without falling out. The key issue with this that medium is that, over time, it will decompose and shed tiny particles that can mess up your pump or drip emitters. Sphagnum is typically bought in flat, compact blocks, and needs to be soaked for around an hour before use

Rice rushes

Rice hulls are a lesser-known and underused substratum in most parts of the world but have proven to be as successful as perlite for crop growth. Rice hulls are a rice production by-product and have the potential to be a cost-efficient, efficient medium in rice production areas.

This free-draining substratum has low to moderate water-holding capacity, a slow decomposition rate, and low nutrient rates. They are not pre-sterilized, though, as rice hulls are a by-product. Growers need to be vigilant using hulls not kept outside or exposed. Rice hulls tend to build up salt and decompose after one or two crops, so they should be regularly replaced.

They are growing Polyurethane Slabs. Polyurethane Increasing Slabs are a relatively new media specifically designed for hydroponic growth. This media consists of about 75-80 percent of air space and 15 percent water holding capacity. As this substrate is so fresh, there is very little information on it.

3.5 Nutrition solution that helps

Various standard nutrient solutions exist, such as Hoagland (1933), Steiner (1961), Bollard (1966), and others. As a general guideline, these standard solutions are excellent but are not suited to different riding conditions.

And if you want to use one of the traditional nutrient solutions, be sure to use the concentration of nutrients as a guideline and not the actual fertilizer formula. The initial composition of the raw water you are using will affect the specific nutrients to be applied with fertilizers.

The electric conductivity is a measure of the total salts dissolved in the solution of the hydroponic nutrient. It is used to monitor fertilizer applications. Remember that EC reading doesn't give you details about the exact nutrient solution mineral material.

The hydroponic nutrient solution is recirculated in closed hydroponic systems and elements that are not absorbed in any high quantities by the plants (such as sodium, chloride, fluoride, etc.) or ions released by the plant build up in the hydroponic nutrient solution.

There is a need, in this case, for more knowledge about the quality of the nutrient solution, which EC cannot provide. Checking the hydroponic nutrient solution will also help you determine when to substitute the nutrient solution, or dilute it with fresh water.

Chapter 4: Hydroponic Nutrients

4.1 What are Hydroponic Nutrients?

The term hydroponic nutrients apply to any commercially available plant nutrients appropriate for hydroponic plants. Plants need nitrogen, phosphorous, and potassium (N-P-K) to develop and various trace elements. Hydroponic nutrients are liquid forms of N-P-K with varying concentrations of secondary nutrients and macro-elements, including trace minerals.

The secondary nutrients are typical of varying sulfur, calcium, and magnesium levels. The liquid nature of the hydroponic nutrients ensures they can be added directly to the root system of the plant. Within a hydroponic nutrient tank, the nutrients are combined with water and added several times a day to the plant's root system. Many hydroponic nutrients also come in the form of powder, but most of the hydroponic nutrients are still liquid.

Due to the soilless growing conditions, hydroponic systems require different hydroponic nutrients. In mediums such as hydro corn, puffed rock, expanded clay pellets, Rockwool, coco coir, Grow stone, grow rocks, or perlite, plants are grown hydroponically sit in.

For deep water culture (DWC), no medium around the roots of the plant is used at all. The plant has no means of consuming nutrients without soil; instead, it must rely on hydroponic nutrients to fulfill its nutritional and mineral needs.

Hydroponically grown plants require to develop regular intakes of hydroponic nutrient solutions. The nutrient-rich liquid solution is distributed by an ebb-and-flow tray into the plant's root system. The hydroponic nutrients are filled several times a day in the tray around the roots of the plant and then drained away into a reservoir for later re-use. When the liquid flows away from the roots of the plant, it is capable of absorbing oxygen.

Hydroponic nutrients are also sold in bottles with two sections, one for growth and one for bloom. That is because plants during their growth cycles need varying quantities of various nutrients. In this way, a grower will dial in the nutritional needs of their crop to optimize yields.

4.2 Formulas of Hydroponic Nutrients

Any grower who either has or is considering a hydroponic system can see that the nutrient mixes are costly. This is a standpoint that leads many growers to make their own, which, luckily, sounds much harder than it is actually.

You will need to understand the fundamentals of what goes into those nutrients and how each part functions before even mixing a single drop. It is because a nutrient solution contains several compounds, which are needed to be made available to your plants during various stages of their development.

Although it is not possible for a total beginner to try, and you will save a lot of money on nutrients. What you need is plenty of planning and a bit of detail. Here, we'll look at what goes into nutrients that enable them to function, and some of the best homemade recipes you can find to substitute purchased nutrients, as well as additional fertilizers and promote development.

Basics of combining hydroponic nutrients In essence, a plant doesn't matter where it gets its nutrients from, whether it's man-made, organic, or something you grow at home. What they think about is they are getting what they need to grow to their full potential. Plants can be picky when grown in soil and eat what they actually want. Yet, in a hydroponic system, it is up to the grower to ensure that these nutrients are available in the right amounts.

Macronutrients are necessary to grow in every single plant. The required ratios of these, however, will be very different in the types of plants you produce. Such homemade formulae have several variations, so making one batch of one kind will produce a different ratio to the next. Hence, we can make a nutrient solution from nutrient salts; it can be easier to produce because depending on the weights of the salts you add, you can fine-tune the blend. When you use these, you must keep these salts cold and dry because any absorbed moisture will affect their weight. Another thing to remember is that many nutrient solutions come in either 2- or 3-part tubes, so you'll look at making either a batch of two or three solutions with any of the following ways while producing your own.

Another thing to remember is that some of the formulations involve measuring spoons. There are yet more things to include, and these are a decent set of weighing scales and rubber gloves for the crystalline chemicals.

In the end, purchased nutrients always come with pH buffers added. Because you're making your own, you're going to need a digital pH-measuring pen and pH-up solution, and pH-down solution. You can find your EC levels out of balance when checking your pH levels, so another gadget you'll need is an EC meter.

Hydroponic Nutrient Mix Formula # 1

This is a one-part blend, known to produce good results. The thing that is needed is to keep an eye on your plants, however, to make sure they display no signs of shortages or burning nutrients.

The formula is appropriate for a water-filled 5-gallon jar.

One thing to remember is that this solution is for non-circulating processes because they are no longer filtered before being applied. This makes them perfect for small systems where you have your origins, like DWC or raft systems, in the solution. If you want to use this in a circulating device, the amounts would need to be increased before you have enough solution to fill your tank.

- Master blend
- Tomato: 10 g
- Calcium Nitrate: 10 g
- Epsom Salt: 5 g

It can be one of the simplest and takes very little time to produce. This solution would need to be disposed of depending on your plants as you harvest from your systems as the levels of salt / EC increase. When there are some symptoms of deficiency, one of the later supplementary formulae can be used to provide some extra nutrients.

Hydroponic Nutrient Mix Formula # 2

The DIY nutrient mix requires a few more compounds than the first one, but mixing is still very easy. When combined, you simply add 10 grams of liquid for every gallon of water you've got in your tank.

- Potassium nitrate: 255 g
- Calcium sulfate: 198 g
- Magnesium sulfate: 170 g
- Ammonium sulfate: 43 g
- Mon calcium sulfate: 113 g
- Iron sulfate: 1/2 teaspoon

Like most homemade nutrient mixtures, make sure you have a container that is wide enough to accommodate a gallon of water and add each of these salts one by one and make sure that each is dissolved before taking over the next solution.

It will be highly concentrated, so just add 10 g to your tank for every gallon of water. In addition, all of your pH and EC levels will need to be tested. You should add a couple of the 'Farmers Friend Recipe' or the 'Gift from the Sea' blend to provide a well-balanced package of nutrients for tomato growers.

Formula # 3

Compost Tea Recipe Hydroponic Nutrient Mix It is the first homemade nutrient mix that can be graded as organic. However, it takes slightly more work than others, but if you have spare space, then this can really help your plants. The first thing you'll need is a compost heap, or a composting bin even better. It will be outside in case of contaminants occur.

You should look at using semi-green waste and half-brown waste when making a compost heap. Farm waste includes lawn cuttings, green leaves, kitchen food waste. The brown waste side contains straw or grass, dead leaves, old papers, wood chippings (not glossy magazine paper).

Turn this every few weeks when you have your stack, so that all the materials can break down and the bacteria do their job. Once your compost is ready, just add two wide shovels full to a big 5-gallon bucket. Fill this with water, and for three days, let it steep.

When you have access to water from the tank all the better, if not, seek to stop any water being contaminated with chemicals. Rainwater is a good choice, too, so start harvesting when it rains.

Once you have soaked your mixture for three days, all you need to do is pour out the liquid and rinse it to eliminate any traces of your compost sediment. You should add the soil back to your compost heap. By using this liquid, for every 50 gallons of water in your tank, you will use 1/2 a gallon. Although this is good enough to use alone, any of the next two recipes or any of the homemade fertilizer or growth booster can also be added.

Hydroponic Nutrient Formula # 4

The Farmers Friend By all accounts, that was developed by a conventional farmer who started his hydroponic foray. He has come up with the following formula, with his experience, which has been well established while maintaining organic elements.

- Seeds: 4 lbs
- Agricultural lime: Gypsum 1 lb
- finely ground: 1 lb
- Dolomite lime: 1 lb
- Bone meal: 1
- lb
- Kelp (Seaweed): 1 lb

We can use it dry, just make sure it is not roasted.

What you need to do to combine is to pour the ingredients into 5 gallons of water. Mix these up until the consistency is thin. Not all ingredients can dissolve, so it is recommended to filter before using. When using this nutrient blend, you simply need to add 6-fluid ounces for every 100 gallons of water in your tank. This mixture is suitable for large systems, so you will need to scale it down to match a smaller system, so you've done enough to use it and don't have to stand.

Hydroponic Nutrient Formula # 5 Sea Gift It can be used as a base formula, and one of the growth enhancers or liquid fertilizers may be applied for a boost. Only adding these in small amounts to your system will provide a good boost to plant growth if you see any signs of deficiency.

- Seaweed (Kelp): 6 oz
- Epsom Salt: 5 teaspoons - 1 teaspoon of water per gallon

This formula #5 is very simple to produce

Just take your seaweed, wrap it in cheesecloth and tie it with twine. This prevents sediment from settling in your bath. Add 5 gallons of water to a bucket, then add your bag of seaweed.

Add the five teaspoons of Epsom salts to this sitting in the sun for five days, and you can apply the whole substance to your hydroponic tank, or you can apply it in one-gallon increments.

With other nutrient blends, to be on the safe side, you need to test the EC and pH levels. It is even more important if you add some of the growth enhancers when growing your plants.

Hydroponic Nutrient Formula # 6

The nutrients here constitute a three-part general-purpose nutrient blend, covering the vegetative phase, flowering phase, and fruiting phase portion. There is also a fourth component, adding a combined compound instead of the individual elements.

These formulas are perfect for one gallon of each Nutrient, so you'll need to change the amounts accordingly if you're planning to scale up.

Vegetative Stage Nutrient Formulation

- 6.00 grams - Ca (NO3)
- Calcium Nitrate 2.42 grams
- Magnesium Sulfate 2.09 grams
- Potassium Nitrate 1.39 grams - KH2PO4:
- Monopotassium Phosphate 0.46 grams - K2SO4:
- Potash Sulfate 0.40 grams - 7 percent
- Fe Chelated Trace Elements Stage Nutrient Solution 4.10 grams - Ca (NO3)
- Calcium Stage Nutrient Solution 4.10 grams.

Chelated trace elements 7.00 percent Iron (Fe) 2.00 percent Manganese (Mn) 1.30 percent Boron (B) 0.40 percent Zinc (Zn) 0.10 percent Copper (Cu) 0.06 percent Molybdenum (Mo)

These trace elements must be blended together and combined into a fine powder in a mortar and pestle before they can be blended to the first three mixtures. How to Blend Your Nutrients When you combine the three mixtures in one go, you'll need enough containers for every solution. For the number of gallons you produce, those should be filled with warm water.

Check the pH of your waters and the TDS / PPM before taking the next steps. When you add your compounds, your pH levels will change. Hold your readings, as, after your final reading, you need to find the true focus. Apply these one at a time with your calculated salts for each compound, and allow each to dissolve before adding the next.

When all of your salts have been applied, let it stand until it completely cools. It is going to be for about 2 hours. Check the pH again after it has cooled, and measure it to the initial reading. Change for your plants with pH UP or down until it is in the right range.

They will be made with these mixtures, so they can add to your tank after they've been cooled. You need to read a second EC because you need to dilute these mixtures before you apply them to your tank. Further, this formula is based on the one given by Keith Roberto, author of many books on Hydroponics. With the number of compounds, it may seem complex and maybe something to try until you have more mixing experience.

Homemade fertilizers and growth boosters, although the above-mentioned nutrients go a long way to delivering all the needs of your plants, they at some stage lack in some areas. To those that are readily available to produce from basic ingredients, this is more the case. It's not possible to get the same intensity each time, so these can be supplemented by the following additional fertilizers to give your plants a much-needed boost. All that follows is very easy to produce and will help your plants grow to their full potential.

Egg Shell Calcium Deficiency Fertilizer

When made, supplying extra calcium to plants is a great solution that you get from this preparation. What you have to do is finely break six to eight eggshells in a mortar and pestle. Add 1 1/2 liter of water and a few drops of diluted hydrochloric acid (wear gloves) to this material. Leave this for 24 hours, and then rinse the water to eliminate all shell traces. Check the resulting liquid to ensure that the pH is about 5.0.

It can be used and combined with every other fertilizer during the growth stage of your plants, which is high in nitrogen. When you use it in the flowering process, then add it to high potassium and phosphorous fertilizer. Moreover, you can use this at any time, but once you have added it to your tank, be sure to check your pH levels.

The Banana Tea Potassium Booster

Potassium is used in all stages of plant development. Bananas are known to contain the largest amount of potassium of any natural product. You can see how important potassium is, even with the earlier formulations, since it is one of the main NPK compounds.

Potassium helps the plants take advantage of the sugars, starches, and carbohydrates that they consume. It helps to create energy reserves, which contributes to the development of complex carbs that give stems and leaves to plants their structure.

Banana tea can also supply essential amino acids which your plants can take up. Only boil three or four banana skins in a liter of water to produce this. You may add a small amount of sugar or molasses to this, which is recommended. Let the solution cool off after heating, and remove the skins. You can do this in the flowering stage of your plant growth, and depending on your plant types; you can gain an additional 20 percent production.

Ground Coffee Growth Enhancer

While you're not going to use the coffee ground on your own, you can use it to make a tea that's a great addition to your plants 'growth stages. One of the key reasons that the resulting tea is so beneficial is that it is mildly acidic, and the acetic bacteria that live in the waste coffee grounds produce 2% nitrogen (NPK) and plenty of other organic nutrients.

You can use these in two ways

Just to add, the first is your composting bin and add the benefits to the resulting liquid, or you can take the simpler route and allow the 24-hour coffee grounds to soak in water. You can use the excess water to add to your tank, and you can then add the waste coffee grounds to your compost bin.

The yeast advantages of Brewer's Yeast Multipurpose Fertilizer Brewer are proclaimed for humans, but it can also be amazing as a plant fertilizer for multipurpose purposes.

One thing to be careful about is that you use brewer's yeast and don't bake yeast since the two are very different.

Exactly what you need to do to prepare this is to add one small spoon of yeast into a liter of water. When this dissolves in a short period of time, it transforms into a natural fertilizer rich in potassium and phosphorus. It can be one of the easiest fertilizers to make, and it can give plants that need a little boost a great helping hand.

Bean Tea Beans and Auxins are abundant in lentils. This helps with the root growth as well as the quest for light for leaves and stems. This, in turn, helps your plants grow taller as they try to meet your rising lights.

For years, these auxins were used in a gel to assist the root growth of seedlings and cuttings. Although these were synthetic, organic products can be just as easily. Many beans and lentils are rich in these, and it is very easy to prepare a bean or lentil tea, which can ultimately extract these. Soak your beans in water until hydrated to the max. You should heat up gently to help the compounds extract.

When washed, whisk, or mix until you have a fine paste, Strain the paste until you have a liquid rich in water nutrition, which contains a lot of axons. It is suitable for the growth encouraged by cuttings or roots. Homemade Nutrient Mixes tweaking because plants are not in the soil; they consume everything they need at the ends of the roots through the minute hairs. By principle, when grown in a hydroponic device, it makes it impossible to overfeed the plants. However, when mixtures are concentrated in too high a nutrient concentration. Your plants won't have enough water to drink.

The salts need to be diluted, so if the concentration of your mixture is too high, your plant will start taking over a water shield instead of absorbing it. Further, this contributes to dehydration of your plants, as the salts suck the moisture from your plants. You need to do so with some patience and caution when you start changing formulae. If you get it wrong, then you will ruin the entire crop. Below are some of the more common nutrient deficiency symptoms in your hydroponic plants.

Nitrogen deficiency: It causes stunted plants with large root systems. The leaves are going to be smaller and a light color. Growth is expected to be sluggish.

Phosphorus deficiency: This leads to stunted plants with small, dull, and discolored leaves. Stems should be unusually strong and have weak root structure. You'll see very little branching off too.

Potassium shortage: Older leaves can turn yellow and curl. The new leaves are dropping, as they get bigger. Flowers will lose luster, and the stems of the plant will be tender and unable to provide maximum support.

Calcium shortage: It allows roots to grow beneath, and you would have curled edges of the vine. Manganese deficiency: This results in slow growth and poor bloom.

Using Homemade Nutrient Mixes All the nutrients mentioned above can be a cost-effective way to add nutrients to your hydroponic system. But when you start growing them, you will need to pay careful attention to your plants.

If you use the formulae that use powdered compounds, each time you end up with a different intensity. In certain cases, this can leave your plants deficient, but this may be rare. If you add too much, the powdered formulas will cause nutrient burning, so these are the ones you need to pay close attention to.

Further, the other downside to purchasing the powdered components is the size of bags they often go into. They will last longer and must be kept as dry as possible. Some moisture will ruin a sack of those minerals quickly and turn them into a big solid block.

On the other hand, you have the handmade organic ones, and particularly the 'Gift from the sea.' If it stays for long periods, this can give off a strong odor. That being said, by making your own hydroponic nutrients, you will save a small amount, and each time you make them, you can learn a little bit more about how to make them, and use them more effectively.

The salt build-up is two of the most important things that you need to look out for. In some systems, you'll see this as the compounds start crystallizing on the sides of your pots or on your rising medium. When flushing the machine, make sure that it is as clean as possible before re-filling the tank.

The last thing to be careful of in the solution is sand. This may not be obvious when you first add your nutrients to your tank, but in a circulating environment, these will be drawn over time toward your pump. You can purchase cheap water pump bags to avoid clogged pumps, which will keep your pump clean without blocking any water flow.

Making your own nutrients can be both enjoyable and beneficial. As mentioned right from the start, your plants don't matter where they get the nutrients from. We will be happy as long as we obtain what they need and grow to their full potential. When you understand your plant types, and what nutrients they are looking for, all these formulas will produce bountiful yields. While some are supposed to be an all-in-one combination of nutrients, this means you may need to inherit something extra in the flowering stages.

4.3 Pros of using Hydroponic Nutrients

Hydroponics is a technique for growing soilless plants, using only water, a nutrient solution, and a structure for keeping the plants up. While diverse forms of water culture have been practiced for several thousand years, the science behind Hydroponics has been more fully understood only in the last 100 years. This has helped both domestic and commercial growers to grow plants in new ways with particular advantages.

No soils required

In a sense, in places where the soil is poor, does not exist, or is heavily polluted, you can grow cultivated. Hydroponics was widely used in the 1940s to supply fresh vegetables for troops in Wake Island, a Pan American airline refueling stop. It is a small region of Pacific Ocean arable. Hydroponics has also been regarded by NASA as the potential farming to grow food for astronauts in space (where there is no soil).

Make better use of space and location

As all that plants need is given and maintained in a system, you can develop in your small house, or the spare bedrooms as long as you have some space. The roots of plants typically grow and spread in search of food, and in soil, oxygen.

It is this Hydroponics, where the roots are sunk in a tank full of oxygenated nutrient solution and close contact with essential minerals, this is not so. This means you can grow far closer to your plants, and thereby save huge energy.

Climate control

As in the greenhouses, hydroponic growers may have complete climate control-temperature, humidity, light intensification, air composition. In that sense, no matter the season, you can grow food all year round. Farmers should grow food at the right time to maximize income for their businesses.

Hydroponics is water-saving

Plants that are grown hydroponically will only use 10 percent of water compared to those grown on the ground. Water is recirculated using this form. Plants will take the water they need, while the run-off ones will be caught and returned to the network. Water loss occurs only in two ways-evaporation and device leaks (but a successful hydroponic design will reduce or have no leaks). Agriculture is expected to use up to 80 percent of ground and surface water in the United States. Though water will become a critical problem in the future as the FAQ predicts that food production will increase by 70 percent, Hydroponics is considered a viable option for large-scale food production.

Efficient nutrient usage

In Hydroponics, you have 100 percent control over the nutrients (foods) plants need. Before planting, growers should test what plants need and how much nutrients they need at different levels, and combine them with water accordingly. These Nutrients are conserved in the tank, and there are no nutrient losses or shifts as they are in the soil.

Solution control pH

All minerals are found in the water. That means that compared to soils, you can calculate and change the pH levels of your water mixture much easier. That ensures the optimum uptake of plant nutrients.

Better growth rate

Are plants growing faster in hydroponic form than in soil? Yeah, it is. You are your own boss who regulates the entire climate for the growth of your plants-temperature, light, moisture, and nutrients in particular. Plants are put in ideal conditions, while nutrients are supplied in adequate amounts, and the root systems come into direct contact. Thus, plants no longer waste precious energy, checking the soil for diluted nutrients. Instead, they move their entire attention to growing and processing fruit.

No weeds

You'll understand how annoying weeds affect your garden when you've worked in the soil. For gardeners, it is one of the most time-consuming activities-till, plow, hoe, etc. Weeds are mainly soil related. So clear the nutrients, and all weed bodies are gone.

Fewer pests & diseases

And like weeds, getting rids of soil helps render your plants less susceptible to soil-borne pests, including birds, gophers, groundhogs, and diseases like the species Fusarium, Pythium, and Rhizoctonia. Even when growing indoors in a closed system, gardeners may easily take care of most surrounding variables.

Less use of insecticide and herbicides

When you don't use soils, so although weeds, pests, so plant diseases are significantly reduced, fewer pesticides are used. That helps you grow healthier and cleaner foods. Insecticide and herbicide cutting is a strong point of Hydroponics as the standards for modern life, and food protection is gradually being put on top of it.

Labor and time savers

In addition to spending fewer hours on tilling, watering, planting, and fumigating weeds and pests, you enjoy saving a lot of time as in

Hydroponics, the growth of plants is known to be higher. Hydroponics has a space in it, as agriculture is expected to be more technology-based.

Hydroponics is a hobby that relieves stress

Having such kind of interest will bring you back in touch with nature. Tired after a hard day of work and traveling, you're back to your tiny corner of the apartment; it's time to lay it all back and play in your hydroponic garden. Reasons such as lack of spaces no longer make sense. You can start your little closets with fresh, delicious vegetables, or essential herbs, and enjoy the relaxing time with your little green spaces.

4.4 Drawbacks of using hydroponic Nutrients

Hydroponics is one of the latest agricultural crazes. Hydroponic crops are plants dressed in a liquid solution rich in nutrients, rather than in soil or other artificial plant bases. Although some greenhouse farmers are going to use media such as wood fibers, clay pellets, peat, or even peanuts packing polystyrene, hydroponic farmers are using something a little safer and more environmentally friendly: water.

With people moving toward more health-conscious lifestyles, demand for crops produced locally has risen significantly over the past decade. Hydroponic farms are one-way farmers now make sure they can satisfy the year-round demand. If you are considering hydroponic agriculture, you are probably wondering what the dangers and benefits of this system entail. Below is a listing of the drawbacks of hydroponic crop production.

A hydroponic garden needs your time and effort

Just like any worthwhile things in life, hard-working and conscientious attitude yields satisfactory results. However, plants can be left alone for days and weeks in soil-borne equivalents, and they still survive in a short time. Mother Nature and soils help to control if anything doesn't fit. In Hydroponics, this is not the case. Plants will die out faster, without proper treatment and awareness. Mind that the future of your plants depends on you. Upon initial deployment, you will take good care of your plants and the device.

You can then automate the whole thing later, but you still need to gage and avoid the operations 'unforeseen problems and do regular maintenance.

Experiences and technical knowledge

You operate a system of many types of equipment that needs specific expertise necessary for the devices used, which plants you can grow, and how they can survive and thrive in a soilless environment. Faults in setting up the growth potential of the systems and plants in this soilless climate, and you end up losing your entire development.

Organic debates

Some heated controversies have emerged about whether or not Hydroponics should be accredited as organic. People doubt whether hydroponically grown plants can get microbiomes as they are in the soil. But people around the world have been growing hydroponic plants for tens of years-lettuces, tomatoes, strawberries, etc., particularly in Australia, Tokyo, the Netherlands, and the United States. They supplied food for millions of people. You can't expect anything in life to be fine. Compared with Hydroponics, there are also more threats of toxins, pests, etc. for soil growing. For Hydroponic growers, there are several suggested organic growing methods.

For example, by using organic growing media such as coco coir, some growers provide plant microbiomes and add worm casting into it. Natural-made nutrients are widely used, such as fish, bones, alfalfas, cotton, neems, etc. Work will continue to be conducted at present and in the near future for this debate on the organic food issue. And then, we'll know the answer.

Water and electricity hazards

Mainly you use water and electricity in a Hydroponic system. Beware of electricity in close proximity to a mixture of water. Working with the water systems and electrical equipment, particularly in commercial greenhouses, always put safety first.

Threats of system failure

You use electricity to control the whole network. So suppose you don't take preventive action for a power outage, the system will immediately stop working, and plants will dry out quickly and die in several hours. A backup power source and plan, particularly for large-scale systems, should therefore always be prepared.

Initial expenses

You are likely to pay less than one hundred to a few hundred dollars (depending on the size of your garden) to buy equipment for your first setup. You'll need containers, lamps, a pump, a timer, a media, and nutrients whatever systems you build). If the system is in operation, only nutrients and electricity (to keep the water system running and lighting) can reduce the cost.

Long Return per Investment

If you follow news about the start-up of agriculture, you might have heard that some new hydroponic indoor company has recently begun. That's also good for the agriculture sector and for Hydroponics production. Even when you start with Hydroponics on a large scale, commercial growers still face some major challenges. Thus this is due in large part to the high initial costs and the long, unpredictable ROI (return on investment).

A simple profitable strategy to encourage investment is not easy to describe, although there are also several other attractive high-tech fields out there that seem to be relatively promising for financing.

Diseases and pests can spread rapidly.

Use water to grow plants in a closed system. In the case of plant diseases or pests, plants on the same nutrient supply will easily escalate to the same. In most cases, in a small home grower's network, diseases and pests aren't that much of a concern. And if you're beginners, don't care much about these things. For large hydroponic greenhouses, it's all complicated. So, best to have a clear strategy for treating the illness in advance and for starters, only using safe, disease-free water sources and through materials, regular monitoring of the systems, etc. If the diseases happen, the contaminated water, Nutrient, and the whole system need to be sterilized quickly.

Chapter 5: Plants to grow

5.1 Handy herbs

If you grow herbs for culinary or medicinal purposes, it doesn't matter; hydroponics is a perfect way to grow them. There are quite a lot of reasons to do so, and the first is that they are rising faster. You can then add to this that they come with more flavor and aroma than counterparts produced in soil do. Research also shows that hydroponic herbs contain more aromatic oils up to 40 percent. Not only this, but growers can grow a variety of herbs that they would otherwise be struggling to grow in their own field.

Hydroponics makes these herbs simple to grow

If we are swinging too low or too high for any of those herbs in either direction, they will end up dying. Growing herbs using hydroponics helps you keep yielding herbs, whatever the weather or season. Hydroponic development takes up only less space and reduces water use. Although all herbs can be simple to grow in a hydroponic system, here are the top eight herbs to cultivate. We're going to go through the basics and benefits of every.

Basil is a common option for hydroponics since this herb is perfect for holding on to the aroma and flavor when used fresh. These qualities are lost on dried basil. So seeing restaurants and greenhouses using a hydroponics device for their basil herbs is not unusual. There are 150 different basil species in total, but the most common ones are: Sweet Basil, Genovese Basil, Thai Sweet Basil, Purple Basil, Lemon Basil, Lettuce Basil, Spicy Basil. Basil can be planted in two ways, by germinating the seeds, or by planting cuttings that form their roots within a week. Basil is a warm-weather herb, so holding temperature that is between 70-80 Fahrenheit is safest.

Blocks of Rockwool are the most common media used in hydroponics with increasing basil. If you can use peat moss, coconut coir, perlite, and vermiculite, these may require sterilization prior to use.

Pythium is a threat to Basil seedlings, you should remember.

What exactly is Pythium?

- Pythium is a fungus that attacks many herbaceous crops and spreads disease. The best way to prevent pythium or other damping-off pathogens is to make sure your growing media surface is not too humid.

- When you get to harvest basil, the top 1/3 to 2/3's of the upper foliage can be cut. The plant will keep growing that back, so you can cut it again.
- Basil will regrow up to 2-3 times before removing the plant altogether, and starting fresh is recommended.

Only cut the quantity of basil you need; this saves the worry of trying to keep it in good shape. Once you pick basil, basil's shelf life is only a few days, so it might be safer to keep it growing on the plant until it becomes essential.

Yet, if we want to see how easy it is to grow perpetual basil in a hydroponic system, you can check this video.

Chamomile

If you're a big tea fan, you may want to learn you can grow your own Chamomile with hydroponics indoors. Chamomile has many impressive antioxidant properties, which have been shown to reduce the risk of diseases such as heart disease and cancer. Often, they help combat insomnia and poor digestive problems.

Many would use a floating seed tray to help the chamomile seeds germinate. You'll want to get rid of the weakest ones after the seedlings grow to around 2 inches, so there's just one strong seedling per cell in the tray. Moreover, it can take up to 1-2 weeks for chamomile seed to germinate.

Chamomile is recommended to receive up to 16 hours of light daily. Chamomile has wide compatibility, as it relates to pH ranges. This can range from 5.6 to 7.5 in anywhere. Ideally, you'll probably want to reach 6.5 in the middle for optimal results to rise. You will be able to harvest your chamomile flowers after about eight weeks.

The flowers can be picked by cutting off up to 3 inches of stem and then drying them in a sunny area on a rug. Through not picking all the flowers, you will make replanting much easier, which helps them to re-seed themselves. For preservation, store your Chamomile in an airtight container in a dark, cool place. You can read more here to see more of the benefits to Chamomile's health.

Rosemary

This Mediterranean herb is an evergreen, with leaves that look like needles. The herb is growing flowers in white, pink, purple, or sometimes blue.

Rosemary can be used as an aid to a wide range of issues such as stomach problems Heartburn lack of appetite Cough, Headache High blood pressure Low blood pressure Toothache Insect repellent And more rosemary growing hydroponically compared to other herbs will prove much slower.

You should predict a harvest time of up to twelve weeks, and the seed yields are often extremely low. They still prove much more effective to grow hydroponically. These herbs are susceptible to infections with the fungus, powdery mildew, and mites. The most suitable for this herb is an NFT hydroponic system, and they should be subjected to temperatures ranging from 70 degrees. However, the max threshold achieved is 85 degrees.

Here are some fast tips to hydroponically cultivate the rosemary. Firstly, Keep the pH range from 5.5-7.0 Moisture levels will remain normal. Secondly, do try to get 11 hours of days light minimum. Moreover, you can harvest 2-3 times per sowing, and this can be achieved during the year.

Oregano

It is a little member of the mint family, and for thousands of years, they have used this herb for cooking and medicinal needs. Yet it was used by the ancient Greeks for treating GI problems, menstrual cramps, urinary tract infections, skin conditions, and dandruff. Several times they have studied Oregano. Oregano for its antimicrobial activity that wards off pathogenic Listeria. Hydroponic Oregano should grow well in pH ranges from 6.0 to 9.0, and the level should fall between 6.0 and 8.0 for optimum performance.

Rockwool cubes are widely used to germinate the seeds that can take anywhere from 1 to 3 weeks. Some other rising media are the Rapid Rooters, Oasis Root Cubes, or Grodan Stonewool.

Oregano is a slow grower, and after a transplant, it can take up to 8 weeks before the first harvest. Oregano likes full sun when you grow outside, and when you grow under the lights, the lighting won't be any different. T5 tubes are suitable for providing the right light, and they should be around 2 to 4 inches from the tops of the plant to prevent drying or burning the leaves.

Cilantro From seed to harvest, when grown hydroponically, you're looking at some 50-55 days for cilantro. This choice of the herb is very low maintenance and needs no trimming. They can be harvested in part or in full.

If you're a food lover, you already know what healthy cilantro is. Toppings, garnishes, salsas, this is what you call it. Although some people don't like the taste, why? A lot of people perceive the cilantro taste differently. Some describe it as a fresh and cool taste, while others consider soap-like to their tastes. Here's a scientific explanation of why this is. A few tips to hydroponically grow cilantro: Keep the pH level between 6.5 and 6.7.

Temperatures of up to 75 degrees Fahrenheit will stay anywhere between 40 degrees Fahrenheit. Nevertheless, for temperatures in the 60s, there are higher germination speeds.

Watch out for spots of powdery mildew and bacterial leaf, popular to cilantro. Such spots cause high humidity levels and exposure to too much humidity. It does need plenty of water, but it doesn't have to be overwatered. Often it is recommended that oscillating air recreate a sturdier outdoor environment.

Anise

This uncommonly heard of the herb has a taste of licorice. Often, it is also called aniseed. Although Anise can fight off many common problems such as digestion, nausea, cramps, and more, other herbs also help. Although the taste of the licorice sort may leave it unpopular with many, it is resourceful in savoring bread, sausages, cookies, and cakes. Anise seedlings are very delicate and difficult to move, so it is best to allow the seeds to germinate and grow without moving them in their respective containers. You'll find the seeds can germinate for up to 2 weeks.

You're going to want to maintain a pH level of about 5.5 to 6.5. Meeting at 6.0 in the middle is the most ideal for development. The seedlings benefit the most from having an oscillating fan stirring the wind gently for a few hours each day.

The best way to harvest Anise is to cut the plant as required, and place it in a protected area free of direct sunlight to dry out. It can hang them upside down, too. They are harvested entirely as soon as the heads start becoming brown. Store away from heat and light in an airtight jar. Anise usually has a shelf-life of up to 1 year.

Dill

Dill is an annual growing herb in the family of celery. It's most frequently seen grown in Eurasia, where it is used to flavor milk. For your recipes, you can use the fresh dill or dried dill. The stems aren't used when using fresh dill. Growing dill is very easy hydroponically, and thrives in this sort of growing climate.

Hydroponic growth tips: Place the seeds on a piece of Rockwool and press them in. Keep the Rockwool moist with nutrients and water waiting for the seeds to germinate? Germination can take 7-10 days but may occur earlier. You can then grow the Rockwool directly into your hydroponic system after germination. Maintain the pH level from 5.5 to 7.5. Enable enough space to grow, and note that sometimes dill will actually grow as high as three feet. Harvest only by cutting the leafy leaves and removing the stems when brown and ripe seeds appear.

Culinary uses for dill include: Soups Salads Dips Casseroles Pickles Medicinal uses include: Relieving stomach bloat and gas Headaches Cramping Catnip If you have a cat, you may want to cultivate this herb hydroponically mainly for their enjoyment and, of course, to provide yourself with some mild entertainment.

Catnip is not only used for cats, somewhat contrary to common perceptions and the name itself. Catnip has been known since the early 1700s for its capacity to alleviate cramps and indigestion when used in herbal teas. Below are some tips for growing catnip hydroponically indoors: By using leaf-tip cuttings or seeds, you can easily propagate catnip.

Provide up to 5 hours of daylight. We were using good drainage to provide a steady amount of water. Catnip can be susceptible to root rot, so try avoiding an area that is too wet. Look out for the growth of molds, which can occur from too much misting. Eliminate any insect infestations, including aphids, mealy bugs, scale, and whitefly. Don't let that cat get close to your machine!

Is it easier to hydroponically cultivate Herbs?

Like often, when it comes to efficiently growing plants and crops, hydroponic systems arise as top contenders do. Herbs can benefit the most from the capacity of the watering system to obtain a continuous supply of nutrients and oxygen. In a hydroponic climate, on average, herbs grow about 25 percent to 50 percent faster than an outdoor soil climate.

However, some herbs are better off being young. Hydroponic systems offer their customers the opportunity to cultivate fresh herbs for restaurants, supermarkets, and commercial farmers, which allows for greater flavor and cost-efficiency. Remember these advantages of hydroponically growing your herbs: You don't need any soil.

While some that love the naturist appeal of dirtying your hands by gardening out the sunlight, the truth is, some of us prefer not to have to go that direction. Hydroponic growth really requires only some water and clear mediums.

You can get bigger yields and quicker growths. As previously mentioned, in a hydroponic system, you would see 25 to 50 percent faster growth than you would with an outdoor crop. This quick development means you'll be able to yield more in less time—more upkeep. Most hydroponic systems are running on autopilot, so you can only test the pH balance and update the nutrient solution regularly.

Herbs most frequently fall prey to insects and pests. The ownership of an indoor hydroponic device would greatly remove such risks. You'll have more energy to save. On average, hydroponic systems only use up to 10 percent of the water used by outdoor soil plants. The water is continually filtered and reused.

You can monitor the environment. Is your region vulnerable to floods, storms, or even temperatures that are frigid? For an indoor hydroponic system that will always be in a tightly regulated and secure environment, you don't have to think about this. You don't need to use herbicides and insecticides, meaning you can keep your herbs 100% organic and safe from harmful chemicals. Using hydroponic gardening, you'll save a huge amount of space. Systems can be tailored and even vertically installed.

Some say hydroponic gardening helps with stress relief. There's always something to put within your home a part of your outdoor world. Finding another living breathing element close, you may have positive effects on mental health.

Getting into it is an all-around enjoyable sport, what's better than the satisfaction you get from realizing you've grown a plant from start to finish, nurturing it every step of the way? If you have a natural green thumb or not, hydroponics is straightforward, for beginners too!

5.2 Vegetables for daily use

Hydroponics is a booming means of growing domestic production. New gardeners who think of this also wonder what the best crops to grow, which are easy to grow and will produce the best yields are. There are many reasons why people turn to this way of rising, and it doesn't matter if it's because they want to help save the world or cut their grocery bill down on it. Hydroponics is a perfect way of doing this, and more.

Though not all vegetables thrive in a hydroponic climate, many do. So here's the end, nine new hydroponic cultivators will develop into their system. Several are quite simple, while others require a little more effort and space, but they are all worth adding to any hydroponic garden. Here we will look at each of these top nine hydroponic vegetables, and which systems are ideally suited to growing them.

Best Hydroponic Vegetables Lettuce Leaf lettuce makes an excellent hydroponic cultivation choice. In simplest systems, it develops and needs minimal attention. You can harvest the outer leaves from your lettuce as you grow, ensuring you'll end up with a prolonged crop of fresh, crunchy lettuce. As the leaves are removed, the leaves inside will quickly grow to take their place.

Many varieties are available to choose from, and most of them are suitable for growing this way – Tom Thumb Boston Iceberg New York Romaine Buttercrunch Bibb Simpson Waldman's Dark Green. Lettuce is ideal for growing in NFT, DWC, and Ebb and Flow systems. When they get too hot on the temperature, lettuce will bolt and taste bitter.

They are a vegetable from the cold weather and prefer temperatures between 50 and 70 degrees Fahrenheit. Lettuce is also fond of high levels of nitrogen.

Kale

Kale is one of the best vegetables grown due to its health benefits and delicious flavor. This can germinate from seeds and can handle a wide range of temperatures from 45 to 85 degrees Fahrenheit once it starts developing.

It takes about ten weeks from seed to harvest, but like lettuce, you can pluck up to 30 percent of the leaves of the plant. Again, there will be new leaves coming back, so you will prolong the time your crops are in your system. You will cut down on harvesting time to around six weeks if you transplant.

One good thing about kale, when grown indoors, is that it's not attacked by many pests. The primary culprit being the aphids, but they may suffer from powdery mildew.

Curly kale (a popular form sold in grocery stores), Lacinato kale (sweeter and with longer leaves), and Red Russian Kale are the main varieties. This variety has a reddish appearance and is the sweetest you can grow.

Spinach

This is Another cool weather crop; this is ideal for growing along with lettuce and kale. Any temperature above a Fahrenheit of Seventy-five degrees would see the plant suffer. This can be grown from seeds, and many hydroponic growers can refrigerate their seeds for up to three weeks before planting. It produces a more resilient plant, and therefore a better plant. Nevertheless, since they are cool weather plants, they do need around 12 hours of light daily, T5 fluorescent lamps might be the better lighting choice.

You should lower the temperatures when it's nearly time to harvest, as that has the benefit of making the crop sweeter. Growth will slow, however, because of this. It is most of it advisable to go for quality over quantity if we are experiencing a bitter taste. Some systems are ideal for spinach, but just remember to plant them a couple of weeks apart so that you can harvest continuously. These can be a raft network ideal because it can also be for lettuce and kale.

Cucumbers

Growing cucumber can be so rewarding in hydroponics. Such vegetables are fond of the conditions given to them—a bit of warmth, good for nutrients, and plenty of moisture. Growers are amazed at the yields as they quickly become one of the most productive vegetables you can produce.

The optimal temperatures for maximum growth are just beyond the ranges, like the greens that are leafy above. They will expand within a range of 60 to 82 degrees Fahrenheit, saying this. This makes them perfect for growing alongside the two next crops on the list.

Cucumber requires a pH of 5.8 with an EC of 1.8 to 2. Growers can consider seeds costly for a good hybrid strain, but this cost per seed is more than justified when you see what fruits one seed can carry once it's developing.

The worst thing about cucumber growth is that they are plants for winemaking and require trellises. It makes them more suitable for flooding and drainage or other bucket-style structures, where there is plenty of through medium to assist with support.

That being said, coconut coir is one of the best mediums to use so long as the plants are well maintained. Look out for such pests as mites, thrips, whiteflies, and aphids. Such insects want to manipulate the cucumber crops.

Nutritious tomatoes

As farmers move on to tomatoes, it shows that they appreciate their method and plan to progress to the next level. What hydroponics is all about is getting a continuous supply of new tomatoes.

They are a plant with warm weather and prefer temperatures that prefer cucumbers. Nevertheless, they prefer an EC level that begins at two and goes up to 5, so they need to separate some framework to allow tomatoes to grow on their own, yet we can make tomatoes at least with other plants that like this level. The perfect pH is between 5.5 and 6.5, and the Fahrenheit is between 58 and 79 degrees. Most probably the upper end of the scale.

They can be planted from seeds, but cuttings or seedlings are preferred since growing fruiting plants from seeds takes too long. There are several different styles, but the vineyards are common as they are easier to manage and harvest from.

Tomatoes as if cucumbers require trellises so that they can grow upwards, and they will produce a steady stream of fresh fruits that you can part harvest from.

Tomatoes can be affected by various pests and diseases such as spider mites, aphids, mosaic virus, and more. Further, yet one thing that can happen depending on the variety of tomatoes is that they may be susceptible to splits. It is when the tomato's interior is rising faster than the exterior. This sometimes occurs in a short time, as they suck up too much space.

Radishes

While most root vegetables aren't ideal for hydroponic production, radish is different. These are a cool crop from the weather so they can complement the first few plants on the list. They also mature quickly, and it just happens that they are one of the easiest plants to grow.

The pH is bets about 6 to 7, with temperatures ranging from 50 to 65 degrees Fahrenheit. If you grow a longer radish variety, these can withstand a little more heat than the types of short bulbs. EC levels are expected to decline from 1.6 to 2.2. The lighting requirements are limited and require at least 6 hours. Between 8 and 10 hours of light are optimal.

We do not suggest seedlings, because they are best grown from seeds. However, it can take as much as little as three or four weeks from germination to harvest. Add to that; you will reap all the way through the year if you stagger your planting. In hydroponic systems where the temperature hovers between 72-76 degrees Fahrenheit, this cool-weather crop grows excellent. The most common radish concern is that if they are not held mist, they can quickly bolt, and if they're too wet, they can suffer from root rot.

Beans

You can grow nearly any sort of bean in a hydroponic garden. There are hundreds of runners, string, pole beans, and bush beans to choose from, and the most popular are. They are easy to sustain and highly profitable for the work that goes into developing them. Some styles require more effort as they climb / wine plants, so they will need trellises for help.

They are fast germination when developing from seed, and can take less than two weeks. Some varieties can even start in as little as seven days. As they grow and you can see that they have two true leaves, they are the perfect size for going into your garden. However, a drip system is also possible, depending on the device form, while ebb and flow is the better choice. When plants are the bush type, they should be planted about 4-inch apart. Pole beans can be lined up a little wider at about 6 inches apart.

One positive thing about beans is they pollinate themselves. The growing medium should be loose, so hydro ton pebbles, or a combination of perlite and vermiculite, are good options and have many benefits.

Perlite does not impact the levels with a neutral pH, and extended clay pebbles do give the roots adequate moisture and oxygen. This is enough for twelve or thirteen hours of light, and the average temperatures will be between 70 and 80 degrees Fahrenheit. When the temperature drops below 60 or rises above 60, then the plant's pod growth would have a knock-on effect.

Beans don't require many nutrients, so you can have a consistent harvest by spreading them apart when planting. For growing plants, this will arrive in as little as 50 days.

Peppers

Peppers are a perfect addition as they can be grown at any time of the year. Not only this, but growers will experience yields that are far greater than if they are cultivated by traditional means. This ensures that fruits are bigger and better quality as the plants are supplied with what they need to help them to grow to their genetic potential.

Ebb and flow systems are ideally suited for this type of crop, although they can be grown comfortably in others that have a strong base of that media for support. Such plants can grow very large, so they need a further spacing between plants of between 7 and 9 inches. That can limit a pot to just two plants. Lighting needs to be about six to eight inches above the plants and need to be changed when they ripen. This can cause scorching if the bulbs are closer than this, and if farther away, it can affect the yield or potential production.

Lighting will be up to 12 hours a day, and no less than 10 hours. In addition, they'll still need ample overnight hours. Average temperatures have to be between 73 and 80 degrees Fahrenheit, and they are ideal companions to be with tomatoes and cucumbers.

During their development, extra attention is needed where stem buds need pruning since the plants are about 8-inch long. It makes the plant devote its resources to growing fruits than other smaller ones. The pH levels must be between 5.5 and 7, and the EC will be within the 3 to 3.5 range.

Celery

Celery may be one of the hardest vegetables to grow in a hydroponic climate, but that doesn't mean it's impossible. It takes up to two weeks for celery seeds to germinate, which is relatively long relative to other vegetables. A quicker option is to use the celery stalk which you bought from the supermarket. When you take the stalk from the bottom and cut it 2 inches, then put the base in a plate of water at room temperature, it will actually start to grow after only a week. Celery requires a lot of water, so a deep water system will be the best system to choose from. Together with germinating seeds, celery harvesting can take up to a total of four months after planting the seeds.

Celery requires a pH level of 6.5, and the nutrient EC level will be between 1.8 and 2.4. This could be an accompanying plant in a growing room built for lettuce crops and cool weather crops. Temperatures of daylight will range from 58 to 80 degrees Fahrenheit. Lighting is not intense, and they need just about 6 hours a day.

Maturity and harvesting can take a long time, and they can challenge a grower who needs patience but growing this crop can be one of the most rewarding given how costly it can be from the supermarket.

Benefits of Hydroponics to Grow Vegetables

It is highly beneficial to grow vegetables, especially in regions where conditions are insufficient, or those times of the year when nothing grows. Some of the above crops can be grown all year round, or you can grow these when planting some of the many others that aren't on the list in a different growing season.

If you want to plant, there are endless advantages, and here are a few you can see: Larger Yields Hydroponics can't make plants grow larger than their nature enables them to grow, but they can grow to their maximum potential and in much smaller space than they do in the soil. The ability to monitor the levels of nutrients and pH in the water also ensures optimal growth for the vegetables leaving no space for failure.

As we have just seen, if the gardener is in full charge, they can use artificial lighting and warning indoor growing conditions to grow throughout the year. Crops that are out of season are costly when brought in, so it makes all the difference to have them a few steps away from your kitchen! More space Hydroponic systems can be installed almost everywhere. You may be hidden from any natural light indoors or maybe concealed in rural fields, or in a garden greenhouse. We can churn out many more crop harvests, but with a much smaller area, that is possible than if the garden was in the soil.

They can quickly provide more than enough food for a large family when a hydroponic garden is up and running. Although some crops are unsuitable, there may be little need to buy any vegetables ever again. Most farmers start by only growing for consumption, but as they go along, they find that they are increasing and need to start getting rid of vegetables because they produce so many. Family and friends will be thankful for delicious fresh vegetables, but there are the shrewd farmers who are turning their gardens into small home businesses. The above vegetables are just the tip of what hydroponics can bring. It's up to you to choose what you expand, but only the ones above mean you can choose from a wide variety. Begin with these, and as soon as you acquire more information or find that you have that little extra room, you can broaden and tackle herbs, strawberries, or anything else you think is hard to get where you live.

5.3 Fruits with quick production

It is a work of process that allows us to grow fruit and melons all year round. However, regardless of the outside weather, particularly during Minnesota's cold winters. But to be effective, you have to consider the growing conditions that a specific fruit requires to survive and pay attention to them.

Water-loving fruits make your hydroponic garden a good pick. Those include watermelon, cantaloupe, tomatoes, peppers, strawberries, blackberries, raspberries, and raisins. Many indoor gardeners even cultivate other, more exotic species of fruit - including pineapples - successfully.

In a hydroponic greenhouse, even some fruit trees can be grown. Banana trees and dwarf citrus trees are common options, for example, lemons. Through using the correct lighting and nutrients, you can grow fruit even in Minnesota in the dead of winter!

5.4 Decorative Flowers to grow hydroponically

Flower gardeners spend countless hours tilling the soil. This makes flower gardening seem to be too hard, and if it takes too much time, so why would you consider growing plants in a hydroponic system?

There are, in fact, many benefits and benefits flower planting has over soil growth in hydroponics. Returns come even quicker, you can tailor your nutrients to every species of plants, and you have no weeds, insects, and fewer diseases to contend with. This results in flowers growing up to fifty percent faster, and yields are much higher than soil production.

Having this in mind, you can now grow flowers throughout the year, and which can be costly to buy when out of season. You may also have as many displays of cut flora around your home as you like. Until looking at each flower in detail, here are nine of the best flowers in your hydroponic system you can create.

The Peace Lily

It can be one of the quick and easy to care for as an indoor plant. But you need the right conditions to rise. These tropical flowers belong to the family Spathiphyllum and are identifiable by lush dark green leaves and open spreading white flowers. Yes, we can grow these inside a hydroponic system, but the thing we do not like is being over watered. In fact, they can be more tolerant than being around too much water under-watering.

Peace lilies, growing in a hydroponic system, are mostly modified variants where they send out small roots for water absorption. The Peace Lilies grown in small aquariums can be seen here. Most farmers sometimes wait until the leaves display wilting signs before watering; this may prevent watering. This can lead to root rot if they are overwatered, and the plant will suffocate.

Peace Lilies Hydroponically Growing Tips: Peace lilies prefer a 68 F to 80 F temperature range. That will yield optimum production. Clean the leaves and reduce the pest attack. Aphids and mealybugs are the most popular. Be sure to keep your lilies moist in draught free areas. The optimal pH range should be between 5.6 and 6.5. Fun Facts on Peace Lilies Despite the term, the peace lilies are not connected to true lilies. NASA researched the Peace Lilies because of their capacity to purify the soil. They show effectiveness in eliminating formaldehyde, carbon monoxide, and benzene from the environment.

However, all parts of the Peace Lily plant can be toxic because they contain oxalate of calcium. This can cause respiratory and stomach discomfort if consumed in large amounts. Love, Lilies, kids, and pets will be kept hidden.

Indoor varieties can grow up to approximately 16 inches while outdoor varieties can grow up to six feet long. The wax plants are also known as Hoya plants. These are a perennial creeper with bush, shrub, and evergreen. They adorn some lovely star-shaped flowers with leaves that can either be smooth or look like a soft felt when blooming. Hoyas are extremely low maintenance, which is one of the reasons why they are so popular with a houseplant.

Hoyas are another plant not taking to overwater. Hoyas may have flowers 1/4 inch in diameter up to 4 "in diameter, depending on the growing conditions. There are several different choices made by Hoya. Hoya, 200 species to be precise, and they all sport their own special shapes and colors. Hoya is half the fight you wish to develop and cultivate.

Hydroponically Growing Hoya tips: Avoid cutting the long tendrils while pruning, this is where flowers grow. Hoyas need sufficient drainage, so they don't get overwatered accidentally. Similarly, this tropical plant can absorb humidity through the soil, which is what makes it so low in maintenance. Find a device where humidity is used. Hoyas prefer indirect light to the clear. They aren't fond of dark corners or clear light. Hoyas can withstand colder weather at temperatures from 50 degrees F up to 77 F. Yet, maintain a pH range of 5.0-6.5 Interesting facts about Hoya. There are more than 200 different hoya variants; you do not find depicted blue, purple, or violet colors.

Some Hoya species have a Crassulacean Acid Metabolism

Each cluster of flowers can contain up to 40 individual flowers on this plant. Snapdragons Antirrhinum is the botanical name for Snapdragons and means 'like a snout.' This plant is native to Europe and North America. Because of their vivid hues, they became a common favorite, and their flowers when pressed imitate a dragon-like mouth.

Depending on the variety, and growing conditions, mature Snapdragons can grow from 6 inches to 48 inches. This means you may need some support for your plant, and the growing medium will need to hold it strong. Perlite is generally more popular. Even when grown in a soil-based medium, flowers like snapdragons are usually grown indoors until they are moved. It is because they are extremely weather prone.

There are over 18 different snapdragons, all of which have green, yellow, red, white, purple, peach, orange, and bicolored blooming colors. Hydroponically, Tips for Growing

Snapdragons: Snapdragons require sufficient irrigation, but not standing in humid conditions. You'll need to dry your growing medium between cycles. Snapdragons are perennials with an only occasional shade that require full sun. They prefer the pH scale from 6.2 to 7.0. Pleasant Information about Snapdragons Also named the Mouth of Dog, the mouth of Lion, the flower of Dragon, and more depending on where you are.

The flowers and leaves have some anti-inflammatory capabilities. If snapdragons are combined with water, they can help detoxify the blood, and increase the production of urine to clean the body from waste. Sunflowers, daisies, zinnia, and chrysanthemums are closely related to Dahlias Dahlias. We know dahlias as octoploids, meaning they have eight homologous chromosomal sets.

If you develop these in your hydroponic system, you need to be sure they've got plenty of space. You must be at least 12 inches deep because you are planting in a tub. Some varieties need more profoundly to be able to rule these out of the system. Dahlias are growing half as large as they are tall; therefore, lateral space is necessary.

Your rising medium will dry out between watering timetables and track your tank rates. When planted, you'll need a 10-10-10 NPK mix to fertilize.

Hydroponically Growing Dahlias tips: Dahlias need continuous light to grow and bloom. They're advised to earn at least eight hours a day. The dahlias are flowers planted in the spring. That means they want warmer temperatures. Stick to temperatures of at least 60 degrees and a maximum of 72 degrees. If you want to have your dahlia shorter and bushier, cut the center shoot above the third set of leaves to promote shorter growth, dahlia requires a 6.5-7.5 pH level range. Pleasant Information about

Dahlias More than 20,000 Dahlia cultivars exist. It's Seattle's official flower, WA though they're not native to the area. In the 18th century, dahlias were known as vegetables but later became more popular as flowers.

They named them Anders Dhal after a botanist from the 18th century. In every way, the Rex Begonias is special. For their flowers, we don't know them, but their leaves and foliage. It is vivid and can be superimposed on dense fibrous leaves. A variety of color variations can be found from brunette, lavender, blue, white, pink, and red. Rex Begonias was first reported in 1856 when the enigmatic plant which no one could recognize was included in a shipment of orchids to England. The plant is tropical and originates in South America, Africa, and Southeast Asia. There are over 1,831 begonia species, each with its own intricate and original appearance on the leaf.

Hydroponically Tips for Growing Rex Begonias: Because Rex Begonias are tropical and subtropical plants, they are native to area types of jungles. That is what makes them perfect as houseplants, as they require little light and prefer shaded and cool areas. These plants can grow to 24 inches in height, from 12 inches. With leaves growing 4-5 inches on average, it's also important to be mindful of increasing external space requirements. When the Begonia Rex has earned too much sun, you can tell, because the leaves will turn brown. Stick to temperatures of 60 to 85 F. For optimum growth, maintain a pH range of around 5.7 to 6.2.

Rex Begonias prefer humid conditions; this may mean their leaves are misting early in the day.

Pleasant Information about Rex Begonias

You can use stem cuttings to propagate begonias. The Begonia stem is specifically built to store water in a way that keeps the Begonia hydrated during dry periods of the year.

On average, begonias have only a life cycle of about 2 to 3 years. When they live in an outstanding and caring environment, they will live more than a couple of years. Begonia plant juice is thought to alleviate headaches and to act as an eyewash for conjunctivitis. It is one of the most varied common flowers to be grown in hydroponic systems by far. However, they are also one of the most commercially produced flowers due to the number of individuals that have them as decorations for the interior.

Carnation petals have a wonderful fragrance, and when they have a presence, make every room feel welcoming. The leaves can be edible as well and are sweet to the mouth. The preferred method is sometimes to expand from cuttings. When propagating in soil, it can take from seeds between two and three weeks. Hydroponics may have this cycle speeded up. Hydroponically: Ensure that carnations receive 5-8 hours of uninterrupted sunlight every day. Carnations can perform at their peak with a pH of about 6.0. Rockwool starter plugs are suitable for seedlings with a height of up to 4 to 5 inches; we can transplant them at this point. Keep the 65 to 75 F temperature range. Ebb and Flow, Dutch bucket, or DWC systems.

They are ideally suited for the growth of the Carnation. When they grow, they will need help. It is assumed that the Fun Facts about Carnations are native to the Mediterranean region. However, no one knows for certain where their roots lie due to the intensive cultivation over the last 2,000 years. The most popular forms cultivated are annual carnations, border carnations, and perpetual flowering carnations. The Greeks and the Romans used the garlands for carnations. Carnations are a group of bisexuals. This means they have reproductive systems, both male and female, which hinder better growth and development.

Orchids are the most beautiful flowers there are for many people Orchids. Due to their woody-thick roots and bright blooms, they're the favorite of a gardener across the globe. One thing that many lovers of this plant don't know about is, most of these are actually grown in hydroponic systems around the world. The reasons for this are that in tropical climates where these orchids grow, they attach themselves to tree bark or between rocks. As the atmosphere can be hot, they are exposed to ample rainwater. When the rain has ceased, it opens the roots to the air and can consume enough oxygen.

Yet, it adds to this surrounding organic matter which rots, and have a steady nutrient stream. This is the purest type of hydroponics, and it is this that makes growing orchids so simple and satisfying in a proper system. Hydroponically Growing Orchids Tips: Use Hydroton pebbles or similar media that have adequate drainage to allow full airflow to the root systems. Maintain temperatures about 60 to 80 F. Orchids also require high levels of humidity and circulation of air. For orchids, a pH range from 5.5 to 6.5 is optimal.

A yet a 400w high-pressure sodium or metal halide bulb can be used as lighting. When under bright lights, orchids use more water. After your orchids have flowered, they can then be shown around your house. Fun Facts about Orchids The orchid flowers will last for up to 6 months. Of the millions of seeds produced by orchids, less than one percent will become a plant. Orchid seeds are the seeds that do not contain endosperm, which is what they need during germination to provide nutrients. Regardless of that, to achieve germination, they require symbiosis with the fungi. Orchid germination can take up to 15 years at times. We use orchids in the production of perfumes, spices, and medicines.

Petunia

These common flowers in South America can withstand hot climates. They're common in borders and in pots across many gardens. There is an almost infinite variety of colors available, which is one explanation of why they're favorite gardeners. Many of the Petunias you see for sale are hybrids and are specifically grown. They can reach just anywhere ranging from six inches to four feet in height as they grow and have a spread of up to three feet. Which means you'll need help and plenty of space to prevent overcrowding. Hydroponically Growing Petunias tips: Petunias needs at least 5 to six hours of full light to grow at its best. When fertilizing Petunias like an 8-8-8, 10-10-10 or 12-12-12 balanced combination.

In germination, however, your Petunias would prefer warmer temperatures, shifting them from this warm area once they have germinated so they can expand in cooler areas. We prefer a lower temperature range from 57 to 65 degrees Fahrenheit—petunias areas 6.0 to 7.0 pH.

Fun Information about Petunias Petunias is an annual plant, and it takes one year to complete a life cycle. Popular Petunias are edible with a mild, spicy taste. There are four categories in which all Petunias fall under Grandiflora, multiflora, and hedgiflora. Petunia originates from the Brazilian word "Petun," which means tobacco. These two plants are connected to each other and maybe crossbred.

Zinnia

Zinnia is a part of the daisy family and is easy to grow. Being native to the United States of Southwest and South America, they prefer to rise in full light. It is recommended to get the best of these colorful plants for at least six hours of full sun or bright light. When they grow, they can range in height from 4 to 40 inches. Which means you're going to need some help and a growing platform to help their root system. Their wide range of bright colors and the ability to withstand hotter climates make them common in many a garden setting for planting. Next year Zinnias must re-seed himself. If you pick for your hydroponic garden from the many varieties, you'll be better off selecting the more compact varieties. Hydroponically grown Zinnia tips: Hold the temperature range between 74 and 84 degrees Fahrenheit. Lastly, for the temperature to as low as 60 Fahrenheit, they can still grow.

Conclusion

The aim is to decide if a healthier plant will be produced using water and nutrient solutions rather than soil. With no unneeded material molecules impeding the roots of a plant, the nutrients may be absorbed faster, enabling the plant to grow faster and healthier. Because of the use of continuous nutrient and water feeding, hydroponic plants have grown much larger and produced faster leaves than normal soil plants. For this reason, the null hypothesis is dismissed as the evidence does not support its reasoning.

The daily use of fertilizer during the day has helped the plants grow at a controlled and steady rate. The null hypothesis suggested the use of a hydroponic system would have no impact on the plant's growth and health. As this was not valid, the null hypothesis was discarded, and the hypothesis of the study is accepted.

Hydroponics is gaining momentum and popularity rapidly, as the best way to produce anything from flowers and food to medicine. Hydroponics is now widely embraced by customers in Europe and is catching up quickly in other countries around the world. You will now be on the right track to grow your first hydroponic crop.

It is basically the method of growing plants without using soil, which may sound counterintuitive to those unfamiliar with the technique. Yet The word itself is an amalgamation of two Greek words: hydro, which means water and protein, which means toil. Plants are rooted in a number of compounds like vermiculite, Rockwool, or clay pellets-inert substances that do not add any contaminants into the atmosphere of the plant. The plant is then fed with nutrient-enriched water.

Hydroponics offers one special advantage over traditional methods of growing. Plants can be encouraged to grow faster by carefully controlling and managing the growing environment, including the amount of water, the pH levels, and the combination of different nutrients. The temperatures of air and soil may also be carefully controlled, as may the prevalence of pests and diseases. The net result is increased yield and greater resource utilization. A less costly approach to resource use means reduced waste, water stock protection, and reduced reliance. These all depend on pesticides, fertilizers, and other potentially harmful substances.

The point in the supply chain where food appears to get wasted most varies from developed countries to developing ones. Losses and wastes continue to occur in developed countries at earlier stages of the food supply chain. Reasons for this include planting, crop management, and harvesting constraints exacerbated by a lack of finance and expertise.

Improving food infrastructure and logistics in developing countries may help solve many of these challenges. Perhaps less unexpectedly, food is usually lost later on in the cycle in higher-income countries. It is also influenced by customer behavior and retailers' approach to discounting practices in-store; sales that do not encourage purchases as food reaches the end of its "eat-by" period inevitably result in waste and loss. The situation is further complicated by inadequate approaches to take unsold food and find other places for it - such as homeless shelters, but not limited to them.

You cannot buy Happiness, but still, you can grow plants, that's pretty much the same thing!!!

References

- ThoughtCo. 2020. *The Secret Innovations And Inventions Of Ancient Farmers.* [online] Available at: <https://www.thoughtco.com/ancient-farming-concepts-techniques-171877>.
- Cosgrove, C., 2020. *Introduction To Hydroponic Farming | Blogging Hub.* [online] Blogging Hub. Available at: <https://www.cleantechloops.com/hydroponic-farming/>.
- Hydroponics?, W., 2020. *What Are The Advantages Of Hydroponics? - Smart Garden Guide.* [online] Smart Garden Guide. Available at: <https://smartgardenguide.com/what-are-the-advantages-of-hydroponics/>.
- Ijser.org. 2020. [online] Available at: <https://www.ijser.org/researchpaper/IoT-based-Automated-Hydroponics-System.pdf>.
- Jason's Indoor Guide. 2020. *Choosing A Homemade Hydroponics Design For Your Homemade System.* [online] Available at: <https://www.jasons-indoor-guide-to-organic-and-hydroponics-gardening.com/homemade-hydroponics-design.html>.

- Library, S., 2020. *Hydroponic Nutrient Solutions*. [online] Smart Fertilizer. Available at: <https://www.smart-fertilizer.com/articles/hydroponic-nutrient-solutions/>.
- articles », F. and Farming, A., 2020. *Advantages And Disadvantages Of Hydroponics Farming*. [online] FabulousDecors.com. Available at: <https://fabulousdecors.com/advantages-and-disadvantages-of-hydroponics-farming/>.
- Maximumyield.com. 2020. *What Is Hydroponic Nutrients? - Definition From Maximumyield*. [online] Available at: <https://www.maximumyield.com/definition/3191/hydroponic-nutrients>.
- Thehydroponicsplanet.com. 2020. *8 Easy Herbs To Grow In Hydroponics (With Pictures)*. [online] Available at: <https://thehydroponicsplanet.com/8-easy-herbs-to-grow-in-hydroponics-with-pictures/>.
- Interiorgardens.com. 2020. *Grow Hydroponic Vegetables Herbs Fruits MN*. [online] Available at: <https://www.interiorgardens.com/hydroponic-plants>.